... AND SO ON

NEW DESIGNS FOR TEACHING MATHEMATICS

John V. Trivett

Burnaby, British Columbia

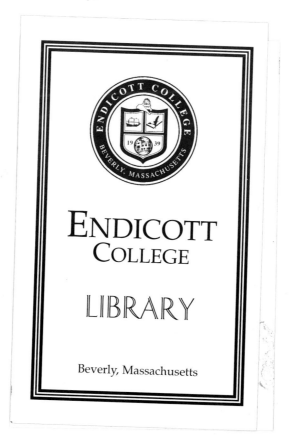

John V. Trivett
Professor of Education
Simon Fraser University

Canadian Cataloguing in Publication Data

Trivett, John V.
. . . And So On: New Designs for Teaching Mathematics

ISBN 0-920490-10-7
    1. Mathematics — Study and teaching
I. Title
QA11.T75   510¹.7   C80-091038-9

© 1980 by Detselig Enterprises Limited
        6147 Dalmarnock Cr. N.W.
        Calgary, Alberta   T3A 1H3

Printed in Canada

Contents

Introduction

The title of this book, . . . *And So On* ("dot, dot, dot and so on") is a phrase common to mathematics. It denotes that what precedes it expresses a pattern, the outer form of some mathematical structure. It also implies that what follows is of the same pattern and that it probably does so endlessly. The example of 1, 2, 3, 4, . . . and so on, is accepted by everybody.

Here the phrase is also taken to refer to other patterns human beings can't help bringing to the learning of mathematics any more than to anything else: advantages of experience, will, insight and ability. Any teacher, therefore, has a responsibility to recognize and have used, not only the subject matter patterns but those of the other kinds, in order that all students harmonize the gifts they have with those being newly acquired – if not endlessly, certainly for the needs and pleasures of many years.

All people, no matter how young or how old, and regardless of outer appearances and behaviour, exhibit patterns based on structure. Because of this and because much of that structure can be interpreted as essentially mathematical – abstractions, relations, transformations, correspondences – radically new approaches to learning and teaching in formal ways are needed to raise significantly the present level of the mathematics of school students and the public generally.

This book is the first of several which will consider aspects of the problem of changing what traditionally has been done. Briefly, there are three ways of looking at it. One, every teacher must surely know thoroughly the mathematics he/she purports to teach, from the kindergarten teacher, a specialist in prenumber work and the inter-relationships of mathematics to language learning, to the specialist of 11th and 12th grades who deals with trigonometry, algebraic geometry and the calculus. Second, a teacher must know the students he teaches are capable of learning that which is offered, learning not merely the names of numbers and shapes or by superficial mimicry, but in the sense of continually becoming the creative owner of the various topics and their connections with each other.

These two requisites are not sufficient, however, for the art of communication is vital, whether it be between teacher and students or between student and student. Unfortunately it is still believed this should be done almost entirely by explanation in the written and spoken language, the students having to fit into what is projected by adult or text. The sense of truth which each student possesses and can exercise, her/his great desire to understand herself or himself – such are rarely capitalized upon, the teachers' or the mathematicians' authority being considered dominant.

It is becoming increasingly clear that efficient communications deeper than surface level, especially those of childrens' inner experiences, their ideas, thoughts, images and fantasies, are much more difficult to maintain. This includes mathematical thoughts – provided we wish students to engage in more of the wealth of abstractions than did their grandparents. For that, teachers will

1

need to become aware of discoveries in other fields, from psychology, therapy, from brain research and the human potential movements. And to know what relationships such non-mathematical attributes of humans have to do with what used to be thought of as simple, "Just tell the kids what to do and see they do it!".

The approach to the learning and teaching of mathematics here stresses knowing rather than knowledge, awareness rather than explanation, the creation of one's own 'problems' as more important than a reliance only on those of the textbook. It encourages the teacher to follow the students. It suggests that we cannot afford to ignore what every student already knows or the perceptions he/she actually has, however 'wrong' they may appear. We must question whether dependance on memory is a valid procedure or whether in its place we pay attention to functionning, just as a person walks well even if he does not remember consciously how he does so.

The approach honors the authority of every individual, his intentions and the fact that most true learning cannot be observed. Sometimes in mathematics lessons learners will be seen manipulating physical materials which aid them in discussion of their ideas, notions, concepts and mental operations. At other times a group of students will give the appearance of a traditional class because they use only paper and pencil or chalkboard. But it is not in the materials that mathematical awareness lies. Neither is it in the choice of topics or in some imagined correct order in which the topics should come – though all of these are valid considerations. It may be assisted only by the teachers' art of placing the complexity of every student's mathematics learning as at least as important as his, the teacher's, teaching.

This book concentrates on the form of the mathematics to be presented, though it does not totally exclude the learning processes. They are implicit throughout though mentioned explicitly and briefly only in the early chapters. Most of the topics are those of the traditional curriculum which certainly will stay in the schools for many years yet.

The book stresses that school students do not necessarily adopt easily the sophisticated terminology, symbolism, writings or development which in the past have been the mainstay of the teaching. It is certainly a long - term aim to encompass all that, but it is not necessarily the hallmark of what should happen on the way. Children and adolescents in their growth patterns come only gradually to accept the adult conventions of language, short cuts and custom and it is best they do so by first establishing meanings and understandings. It is precisely the basics of those 'child-like ways' that are so important in the growing of every human being from conception to any and every present moment. They, therefore, need to be seen as precisely the basics of the mathematics learning too. That need not deny the nature of the subject or the requirements of society.

Further books in this series will concentrate on secondary mathematics, on geometry and upon the basics of the learning processes for practical implementation in classroom.

The first three chapters may interest those who feel some 'math anxiety'. Intentionally their topics are not some commonly met in school, though they could be with great effect. Since at first glance the mathematics of subsequent chapters may well remind many of the dull rote learning they themselves experienced, the traditional topics are delayed until readers have had an opportunity to tackle some unusual situations which look like games – yet contain important mathematics. Hopefully by the end of chapter 3 everyone will be convinced that some mathematics is easy and enjoyable. Therefore it may be conceivably so even with whole numbers, fractions and decimals!

However no one is compelled to begin with chapters 1, 2 and 3. Nor need they be sampled for more than a taste before chapter 4 is begun. That discusses close relationships of mathematics to the learning of language, in which everybody has had success. Nevertheless it is not generally recognized that arithmetic is mainly a matter of finding different names for numbers and is in so many ways a language art. In our schools the two subjects, Arithmetic and Reading are separated!

Chapter 5 begins the familiarly entitled topics but with new insights. From there onward the implications are that computation will not be best solved by the coming of computers to the classrooms, but by children developing fluency in the number language; that the wide - spread difficulties with fractions will not be eliminated by the use of the metric system; and that decimals and percentages are only more variations in the names of numbers.

Readers are urged not to read the book like a novel. Important is what he/she does to one side, on paper, in discussion or with the recommended materials. Fingers will, of course, be available and even perhaps some gathered pebbles. Activities with colored rods are also mentioned because they have been in schools for years.

Exercises are provided but the book is not a compendium for classroom implementation or quick reference. Nor is it a recipe book with dozens of possible class activities. Such help is detailed elsewhere. The hope is that readers will take seriously the frequent urge to make up examples *of their own.*

Many past and present beliefs are challenged: the nature of mathematics learning, its teaching and for which group of people the discipline is intended. It is held that mathematics is for *everyone* . Society has recognized this for a century by insisting that the subject occupy an important place for at least ten years in the school life of every youngster. Now we must see to it that every one of them experiences the joy and personally-felt ownership of one of mankind's greatest accomplishments – Mathematics, which, as Le Corbusier said, ". . . is the whole majestic structure by which Man comprehends his Universe."

. . . *And So On* is a contribution toward everybody knowing that and, as one typist had it, . . . And soon!

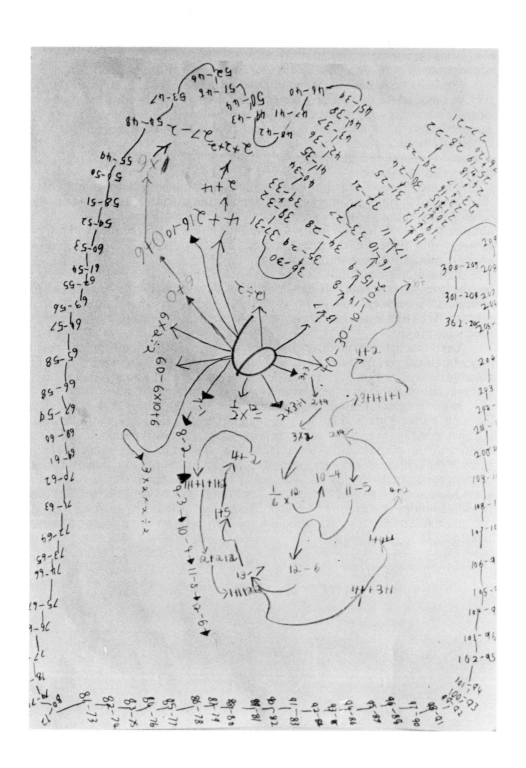

1

My Three Sons – A Game

Listen to a Story.

The Managing Director of General Motors has three sons. Each year on the morning of January 1st he presents each son with a brand new car. Of the three new cars needed every New Year's Day one is a Buick, one a Chevrolet and one a Pontiac.

Exercise 1

Suppose the son's father is interested in different ways these change - overs can occur.
For example, the son who had a year-old Buick could receive a new Pontiac; he who had a Pontiac could get a new Buick and the son who had an old Chevrolet could receive a new Chevrolet. That would be *one* possible change - over.
A change-over always involves 6 cars: 3 old and 3 new.
What other possibilities are there?

You have probably written something on paper to represent the possible change - overs.
Would it be useful, do you think, to use initial letters for the makes of car? B, P, C?
Perhaps an arrow could be used to represent 'changes to' so that something like this is produced:
$B \rightarrow P$ $P \rightarrow B$ $C \rightarrow C$ This shows the possibility of the change - over given in Ex. 1.

Alternative ways of representing
a change over:

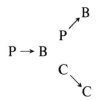

B P C
↓ ↓ ↓
P B C

P → B

(We shall choose one of these diagrams to use, but any and others may be just as useful.)

Are you satisfied that B P C *does* represent one possible change- over?
↓ ↓ ↓
P B C

True, one son does not change the *make* of car, but he *does change* cars and that is what is intended.

Is this another possibility? B C P
↓ ↓ ↓
P C B

It may at first glance look like a second possibility. However, it means that the son who had a B̲ uick changes to a new P̲ ontiac

        the son who had a C̲ hevrolet changes to a new C̲ hevrolet

        the son who had a P̲ ontiac changes to a new B̲ uick.

This is not, however, a different change - over. It is exactly the same as described above; it was just *described* in a different order. If the circular notation is used for the letters and arrows, this might be even clearer, because we might not be influenced by the left to right reading!

Note also that it does not matter which particular son it is who has a particular car, only 'he who had a Buick changes to a new Pontiac . . . and so on'.

Have you then finally decided there are six possible 'change - overs'?

B P C   B P C   P B C   B P C   C P B   C B P
↓ ↓ ↓   ↓ ↓ ↓   ↓ ↓ ↓   ↓ ↓ ↓   ↓ ↓ ↓   ↓ ↓ ↓
P B C   B P C   C B P   C B P   B P C   B P C
  1   2   3   4   5   6

For the sake of naming these 'change - overs' let us number them as shown but, of course, they could be numbered otherwise.

Exercise 2

Suppose a change - over occurs for 1970/71 and another for 1971/72.

Is there *one* change - over which could transform the 1970 state of affairs to that of 1972?

Or, in other words, what *pair* of change - overs is in a sense equivalent to a *single* change - over?

(It is not equivalent, for example, in the time they take.)

The kind of discussion for Exercise 2 may go something like this:

number (3) change - over   P B C
(see above) is: ↓ ↓ | ↓
                         C B P

PBC *could* be the 1970 state of affairs.

So CBP is now the 1971 state.

If next time change - over (5) is used, then C → B, B → C, P → P according to (5) above.

Summarizing:

```
P   B   C   )                    P   B   C   )
↓   ↓   ↓   )  This is (3)                   )
C   B   P                                    )  This is (4)
↓   ↓   ↓   )  This is (5)                   )
B   C   P   )                    B   C   P   )
```

In each case PBC was transformed into BCP, either in two steps or in one. So (3) followed by (5) is equivalent to (4).

Exercise 3

We have examined (3) followed by (5). We can write $(3)(5) = 4$, $3 \cdot 5 = 4$, or even $(35 = 4)$ so long as we know it means "(3) followed by (5) is equivalent to 4" (and has nothing to do with thirty - five in this context).

What then is the equivalent of $(5)(3)$? of $(1)(6)$ . . . and so on?

In fact, this is an extension of Exercise 2 except to ask now, how many pairs of change - overs are possible, and what is the equivalent of each?

All we have to do is choose a pair: $(5)(3)$ or $(1)(6)$ or $(2)(4)$ and work through each, gradually accumulating the results about which we feel certain.

If the work is tackled by a group of students the total number of pairs can be parcelled out until all have been done.

A useful way of recording the results is by means of a table:

	(1)	(2)	(3)	(4)	(5)	(6)
(1)	.	.	.	.	.	.
(2)	.	.	.	.	.	.
(3)	.	.	.	.	4	.
(4)	.	.	.	.	.	.
(5)	.	.	6	.	.	.
(6)	.	.	.	.	.	.

Two results, $(3)(5) = (4)$ and $(5)(3) = 6$, have already been inserted. There are 36 altogether.

Exercise 4

Complete all the entries in the above table.

Before proceeding it will be wise to check the table carefully. As more and more entries became available, did certain patterns arise which acted as a guide to completion?

For example, if it was found that $(5)(3)$ was equivalent to (6), was it true that $(3)(5)$ also gave (6)? In other words, could one reverse the order of a pair of change - overs without affecting the entry?

Further, is the pattern of the table such that a number is repeated in any line or in any column? Or does each number only appear once in every row and once in every column?

Do you agree with this completed table? (we can leave out the parentheses if there is no danger of misinterpretation).

	1	2	3	4	5	6
1	2	1	4	3	6	5
2	1	2	3	4	5	6
3	6	3	2	5	4	1
4	5	4	1	6	3	2
5	4	5	6	1	2	3
6	3	6	5	2	1	4

Exercise 5

What other patterns do you notice?

Exercise 6

With the patterns discovered so far, which of them depend upon the way the table is organized?
   For instance:
a) If instead of using 1, 2 ... 6 we used A, B ... F to identify the various change - overs, would a pair and its reverse be equivalent?
   Would AB = BA?
   Would this ever be true, sometimes, always or never?
b) If the table above is arranged so that the numbers in the headings were in a different order, what would we see?

Exercise 7

Choose *any* order for the change - overs, say 1, 3, 2, 6, 5, 4.
Make this the order of the horizontal heading and the vertical heading. *Rewrite the table* by copying the answers from the first table. The entries will be in different places of course. Try a few different tables like this. Inspect each table for patterns not seen previously.

Here are two other tables, *equivalent* tables in the sense that every entry is exactly as before, but the arrangement of the entries is different:

	1	2	3	4	5	6
1	2	1	4	3	6	5
2	1	2	3	4	5	6
3	6	3	2	5	4	1
6	3	6	5	2	1	4
5	4	5	6	1	2	3
4	5	4	1	6	3	2

	2	4	6	1	3	5
2	2	4	6	1	3	5
4	4	6	2	5	1	3
6	6	2	4	3	5	1
1	1	3	5	2	4	6
3	3	5	1	6	2	4
5	5	1	3	4	6	2

Look at the entries in each quarter of these tables. We shall have occasion to refer back to them later.
   By this time it is most likely that everyone will have noticed that whatever table is used, one of the rows is always exactly the same as the horizontal heading and one column exactly like the vertical heading. In the numbering we began with in this story it is always the row and column headed by (2).

Change - over (2) does not affect the state of affairs at all. For each row, change - over (2) means each son certainly gets a new car but there is no change in the make.
   Change - over (2) is called a 'neutral element' in the system as shown in the table. It is the *only* neutral element.

Exercise 8

Using one of the tables, extract the pairs of change - overs which are each equivalent to the single change - over (2).
How many of them are there in all? Will this correspond to the number of (2) entries in the table?
More of this later.

The story continues:

In constructing a table we paired two change - overs together and found that every pair is equivalent, in a sense, to *one* change - over. Not in all senses, because two change - overs occur only over a year and a day while one change - over takes places only over 1 day. Or maybe we should say 1 year and an instant, and one instant, since a change - over takes place as an instantaneous decision made by father!
Suppose we now study *three* change - overs successively? Would this give us interesting patterns?
Let's consider one possibility, say (1) (3) (5), in that order.
What could this mean?

Well, we could refer to the table for (1) (3). This gives (4).
So for (1) (3) we substitute (4)
(1) (3) (5) = (4) (5)

Now for (4) (5) the table gives 3.
Finally, therefore, we can say
(1)(3)(5) = (4)(5)
= (3)

Exercise 9

Work through (1) (3) (5) in the same way, by first entering (3) (5) in the table, and then from the table again using (1) followed by the single result of (3) (5).

Exercise 10

Choose any other three change - overs. Condense each to one equivalent change - over by the *two* methods above, viz.,
First: Associate the second with the first;
    obtain its result from the table;
    associate this with the third change - over;
    obtain this result from the table.
Second: Associate the second with the third;
    obtain its result from the table;
    associate the first with this result;
    find in the table the entry corresponding to this.

This is longwinded and can be, perhaps, more easily seen like this:
$$((1)\,(3))(5) \qquad \text{and} \qquad (1)((3)\,(5))$$
Note the order has not changed. It is (1) (3) (5) in both cases.
What do you discover about these two strategies for condensing the three successive change - overs to the equivalent of a single entry?

No matter which three change - overs are chosen the two strategies described above come to the same end. It does not matter, in tackling (1) (3) (5), whether we first associate (1) with (3) and their result with (5), or whether we associate (3) with (5) and then use (1) with this equivalent. Finally, in both cases, for (1) (3) (5) we get (3).

One last exercise is worth mentioning because we need to know whether the pattern just described holds for all cases of three successive change - overs or whether we have by chance chosen one example which works!

If we are confident that the pattern works in all cases by working through them all, we shall be in a more powerful position of mastery of the whole system. Fortunately, as you will soon realize, there are methods of proving all this without having to grind out every possibility.

Exercise 11

How many cases of three successive change - overs are possible?

Summary

This activity can end with the discovery of the patterns as a result of the exercises suggested. Some of the patterns have been identified. They are listed here because they constitute an important topic in modern mathematics which has vital application in statistics, physics and technology:

1. The table shows pairs of elements ('change - overs' in this example) which when combined by an operation ('followed by' here) correspond to single elements. This occurs for *every* possible pair of elements (technically, the operation is 'closed').
2. The operation is 'associative' (for any three successive change - overs there are two equivalent interpretations, depending on which element is first associated with the second one).
3. There is one 'neutral' element, in that when this is combined with any other element by the operation, the single equivalent is the 'any other element'.
4. Some pairs of elements give the neutral element as their single equivalent. Each element of such a pair is called the 'inverse' of the other in the pair. Every element in turn has an inverse.

e.g., (4) is the inverse of (6), because (4) (6) = (2)
(2) is the inverse of (2), because (2) (2) = (2)
So, (2) is its own inverse.
(What are the inverses of (1), (3), (4) and (5)?)

Whenever these four patterns or principles are true for any set of elements with an operation which combines them in some way, we can say the systems have the *same structure* . (A system is made up of elements and an operation.)

Exercise 12

Go back to the last table, the one in which the top left quarter is thus:

	2	4	6
2	2	4	6
4	4	6	2
6	6	2	4

Has this small table within the large table the same structure?

We can test the sameness of the structure by checking each of the four criteria described above:

1. The system of Ex. 12 is indeed closed. There are no blanks in the table, so *every* pair is equivalent to a single element.
2. By testing any set of three elements you will find the associative principle holds true.
   Try (2) (4) (6) and (2) (2) (2) and others.
3. (2) is still the neutral element.
4. What is the inverse of (2)? of (4)? of (6)?

The structure of the small table is the same as that of the large table.

Finally, note that in these systems we cannot always reverse the order of any two elements combined in the operation and get equivalence. Generally, one *cannot* reverse, though there are exceptions in the large table, when the neutral element is one of the operations:

In the large table (4) (3) = (1) but (3) (4) = 5
In the large table (2) (5) = (5) (2)

What happens in the small table?

The Learning Approach

The purpose in using this fanciful story was to provide what was probably a new experience to those who perhaps think of mathematics as beyond them, or who have carried bad attitudes to the subject.

From the point of view of learning how to teach the subject a reader is advised now to study precisely what happened as he progressed through the story and the exercises. If the work was shared by a group there will be a greater opportunity to tease out from the combined experiences, the strategies, difficulties, mistakes, temporary annoyances, frustrations and confusions. But it can be done also, if it has to be, by any individual working on his own.

Questions such as these can be addressed:

1. Did you quickly become clear as to essential rules of the game?
   . . . or did supplementary questions need to be asked?
   . . . or were beginnings made on the problem, to be abandoned subsequently, with new starts made?
2. Did you feel strongly that if only the author had explained more clearly what he wanted, the task would have been easier? . . . or is it that such clarity is seldom quickly attainable, except in the simplest circumstances?

3. Did you forget parts of the story which you vaguely thought may be important?
4. Did you remember parts of the story which gradually were found to be unimportant?
5. Did you need to listen to or read the story only *once* ?
6. Knowing that you would probably forget as you progressed, did you resort to noting something down on paper?
7. Having chosen some form of symbolism and model for working on paper, did you perhaps abandon your first choice – or did others – until you agreed on a scheme shared by the group?
8. Are there some criteria needed so that we can decide which form of symbolism and system of writing are satisfactory for the job in hand? What, for example, about size of paper space used? Are abbreviations of words, their initial letters perhaps, sometimes useful?
   If there are several sets of symbols to be used, is it worthwhile having letters for one, numerals for another, other made - up signs for others, so that ambiguity can be avoided as far as possible?
9. In the construction of a table do patterns gradually emerge and help the constructor to fill in more? . . . could some patterns spotted in this way hinder the correct completion of the table, because the pattern assumed partway is not one which is true for all entries?
10. Does each reader/student pause from time to time while tackling the activity?
    . . . does this necessarily coincide with the reflective periods of other learners?
    . . . nevertheless, in a book, or in a group of people, working together, does not *some* surrender have to be made to what others do?

All such considerations have important impact in learning mathematics, in kindergarten, in college and also in other subjects.

The Mathematics of Chapter 1

The essence of the General Motors story from the standpoint of mathematics, is that it embodies a system with a structure which is exhibited by the table and the patterns inherent in it.

The system consists of 'elements' and an 'operation'. The elements in this story are the 'change - overs' – six of them. The operation is the process by which each of the thirty - six pairs of elements can be associated with just one element.

The story used here is only one possible application of such a system. We shall see later that there are applications to numbers and their operations with the same structure.

Formally, the system is called an 'abstract group', which is a system of elements and some way of combining pairs of them such that these rules are true:

1. The system is **closed** . For *every* pair of elements in the system which are combined, there is a corresponding single element of the same system.
2. The **associative** rule holds good. For every three elements combined there are two equivalent ways of associating them with one element.
3. One element is a **neutral** element. When it is combined with a second element the single equivalent is the second element.
4. Every element has an **inverse** . When an element is combined with its inverse, the single equivalent is the neutral element.

2

The Tower Game

There is a game on the market, sold in stores, called the *Tower Game* or the *Tower of Hanoi*. It is a very ancient game and you are now invited to play it.

It consists of three vertical thin posts sticking out from a base block. Onto the middle post ten circular discs are stacked, each having a hole at the centre. The discs are graduated in size, the smallest being the top of the pile, the largest at the bottom.

The object of the game is to transfer all the discs from the middle post to one of the others, honoring two rules:

             1. only one disc can be moved at a time.

             2. a larger disc must not be placed on top of a smaller one.

Please begin to play the game.

Exercise 1

Since we have not supplied a model of the game, not even a drawing, can you improvise in some way and begin playing?

Do so before reading the next paragraph.

Plateau

A plateau is an opportunity for a pause, perhaps a rest, a looking back at what has happened so far, a brief summary of the essence of the game and a look forward. Take this opportunity when you have had some experience at playing the game, using whatever improvisation you came up with.

If this is done by several people working together, or a teacher leads the activity in class, it will be easy to exercise patience while others improvise and, through conversation, students will be influenced by others.

With a statically printed book as the guide there is a strong temptation to push on before much experimentation has beeen done by the individual reader. We beseech you to try and resist this!

Play the game, therefore, with *your* improvisation, before reading on.

In a group of people playing the game these are some of the reactions which commonly occur.

The students settle down and draw the game on paper. To a keen ear comes evidence that this is being done:

> "Take the top disc, put it there. Note the second disc on that post. Where will the next one go? No, it can't go there because it will be on top of that one which is smaller and that's against the rules. It must go *there* ."
>
> Others may cut out paper discs, placing them on top of each other, inside one of three circles drawn on the table. Or objects may be extracted from pockets and used. Coins may be tried, then discarded because there are not enough of different sizes.
>
> Sometimes students may meanwhile be wondering what to do, saying they don't understand the rules or don't know what to do.
>
> One student may cry, "I've done it," for the rest to discover that he has decided to place each disc separately on the table, then pick them up in the reverse order and place them onto another post! "Anyone can do that," a colleague replies, but the others take the view that if *that* is the game one wishes to play it does no harm, though something more challenging may be preferred!

What now are the drawbacks or advantages of your improvisation?

If drawings, do they become too complicated as you try to draw successive moves?

If objects, do they stack well enough, or can they be laid horizontally on the table, in some definite order, which is always recognizable as equivalent to the vertical arrangements?

Which are preferable, actual objects to move around or drawings?

Did you understand the rules immediately or did they become clear only as you experimented in the game?

Did you think extra rules were needed prior to beginning, or could they be inserted as you went along?

This is the point at which for this chapter we can probably settle on our model. We choose a set of discs marked from the top a, b . . . j (a set of colored rods, the initial letters of their color names being used for identification, could also be used as a physical model).

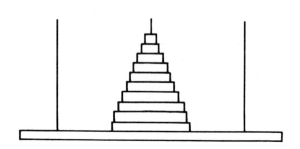

The game continues:

What moves of discs (or rods) are you sure of?
If the discs are initially on the middle post does it matter whether the first disc is moved to the left post or to the right?
If you think it matters, how will it appear to another student sitting facing you on the other side of the model?
Do you agree with the following at least?

Move 1

Disc a   is moved from M (Middle post) to either the left post (L) or to the right (R).
We choose L here, although we could have chosen R.

Move 2

b   is moved to R.

Move 3

a   is moved on top of b.

If there had been only two discs, a and b, this would have concluded|the game, though it would be rather insipid and not much of a challenge. Nevertheless, you are *certain* that you can play the game for one disc, and also for two discs!
You must have *begun* the complete solution.
Can you now experiment with three discs: a, b, and c?

Is this what you find?

Move 1

a → L (now using more shorthand, although this is not the only possibility and you may prefer otherwise).

Move 2

b → R

Move 3

c ↛ L (c cannot be moved to L, because it would cover the smaller a)
c ↛ R
a → R (on top of b, which is larger)

Move 4

c → L

Move 5

a → M

Move 6

b → L

Move 7

a → L and that does it, for the 3 disc tower is now in place on post L!

Plateau

An alternative method of recording:

Table I

L	M	R	
-	abc	-	) Move 1
a	bc	-	} Move 2
a	c	b	) M3 This is halfway
-	c	ab	through the game
c	-	ab	
c	a	b	
bc	a	-	
abc	-	-	

How many *moves* is this, if no unnecessary moves are made?

How many *states* are there represented by each line of description in the above table?

What *patterns* do you notice in the above table?

The game continues:

Try the game for four discs.

As you begin to succeed, ask yourself if in any way the three disc game is used.

If a trial does not work, begin again, and try something else. The recording table can be used as a reference, unless your hand is beginning to move in a pattern of success!

Plateau

Your record may now look something like this.

Table II

	L	M	R	
	-	abcd	-	
a → L	a	bcd	-	
b → R	a	cd	b	
a → R	-	cd	ab	
c → L	c	d	ab	
a → M	c	ad	b	
b → L	bc	ad	-	
a → L	abc	d	-	
d → R	abc	-	d	
a → R	bc	-	ad	(halfway)

b → M	c	b	ad
a → M	c	ab	d
c → R	-	ab	cd
a → L	a	b	cd
b → R	a	-	bcd
a → R	-	-	abcd

Exercise 2

If the four discs were first on R and the first move consisted of moving   a   onto M, what would the rest of the table look like?

L	M	R
-	-	abcd
-	a	bcd

Exercise 3

What is the table for 5 discs?

Exercise 4

What patterns can be extracted from the 3, 4 or 5 - disc table?

Exercise 5

How many moves are needed to succeed in the 3, 4 and 5 - disc games?

Exercise 6

Is the 3 - disc table embedded within the 4 - disc table? The 4 - disc table within the 5 - disc table?

Exercise 7

Do you agree with this indication of the embedding:

Table III

	L	M	R
	-	-	a|b|c|d
1 disc game	-	a	-|b|c|d
	b	a	- -|c|d

2 disc game	ab	-	- - c d
	ab	c	- - - d
	b	c	- - a d
	-	bc	- - a d
3 disc game	-	abc	- - - d
	d	abc	-
	ad	bc	-
	ad	c	b
	d	c	ab
	cd	-	ab
	cd	a	b
	bcd	a	-
4 disc game	abcd	-	-

The game continues:

By now you probably agree that the total number of moves are as follows:

1 disc game	1 move	(or $2 - 1$)
2 disc game	3 moves	$(4 - 1)$
3 disc game	7 moves	$(8 - 1)$
4 disc game	15 moves	$(16 - 1)$
5 disc game	31 moves	$(32 - 1)$

Is it reasonable to anticipate 63 moves next? . . . and so on? These can be alternatively written:

$$2^1 - 1, 2^2 - 1, 2^3 - 1, 2^4 - 1, 2^5 - 1, \text{ and so on. } (2^4 = 2 \times 2 \times 2 \times 2)$$

If we were interested in the total number of 'states' (or disc arrangements) the sequence would be

$$2^1, 2^2, 2^3, 2^4, 2^5, \ldots \text{ and so on.}$$

One question which sometimes arises is the significance of whether the first move of a is to L or to R. Placing the posts in a circle instead of a line produces an alternative insight into what may have been spotted already.

Exercise 8

Play the 4 - disc game with the posts in a circle, beginning at any one of the three discs.

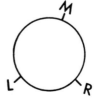

How does disc _a_ move? disc _b_ ? disc _c_ ? etc.

Exercise 9

Return to Table III. Examine how _a moves, then _b , _c , and _d .

The game continues:

Is this the order in which the discs move?
a, b, a, c, a, b, a, d, a, b, a, c, a, b, a
_a was moved every other time, and it moves in *cyclic order* just as it does
when L, M, R are placed in a circle. If it is at R it next moves to M, then to L and
then back to R.
Disc _b moves in the opposite direction R – L – M – R – . . .
Disc _c moves in the same direction as _a ; _d moves as _b .
Generally, it seems as though discs _a , _c , _e . . . will move in one direction
(if there *are* as many discs) and _b , _d , _f , . . . in the opposite direction.
_a will move on the 1st, 3rd, 5th, 7th . . . moves
_b will move on the 2nd, 6th, 10th, 14th . . . moves
_c will move on the 4th, 12th, 20th, 28th . . . moves
_d will move on the 8th, 24th, 40th, 56th . . . moves and so on.

Exercise 10

Complete this table for 10 discs, begun above.

Table IV

a	1,	3,	5,	7
b	2,	6,	10,	14
c	4,	12,	20,	28
d	8,	24,	40,	56
e	16,			
f				
g				
h				
i				
j				

What is the pattern of the first numbers in each row?
What is the pattern in each row for anticipating the numbers in the
sequence?

The game continues:

With this table completed, the game can be played easily and with confidence – almost mechanically.

<u>a</u> moves first and on every other move.   Either direction may be chosen, but once decided <u>a</u> will always move thus.   Particular moves can be predicted. Asked a question like "what move is the 16th move of <u>a</u> ?" we need to find the sixteenth term of the sequence 1, 3, 5, 7 . . .

Term 1 is 1
Term 2 is 3
Term 3 is 5 or $(2 \times 3) - 1$
Term 4 is 7 or $(2 \times 4) - 1$
Term 5 is 9 or $(2 \times 5) - 1$ and so on.

. . . . . .

Term 16 will be $(2 \times 16) - 1 = 31$

The other sequences are slightly more complex but each term can be rewritten and a pattern spotted.

disc b   2, 6, 10, 14 . . .
or: 2, 2 + 4, 2 + (2 x 4), 2 + (3 x 4), . . .

Exercise 11

Write in similar alternative ways the first few terms of the other sequences and use them to say what is the 50th move of each disc (if there *are* enough discs in a game to warrant that number of moves).

The game continues:

The reverse kind of question can also be asked, of the form: "What disc is being moved during the, say 73rd, move of the game?"

In this case, the number 73 is odd. It must, therefore, be a move of disc <u>a</u> and since $73 = (37 \times 2) - 1$ it is disc <u>a</u> 's 37th move.

"What disc is being moved during the 60th move?"

If referring back to Table IV it has been noted that number sequences also appear in *columns* 1, 2, 4, 8, . . . 3, 6, 12, 24, . . . 5, 10, 20, . . . it will be seen that every number, after the first, in every column is *double* the preceding number.

We can use this pattern in considering the 60th move. Dividing 60 by 2 gives the number immediately above it in its column. That is 30. $30 \div 2 = 15$.

15, being odd, is in the <u>a</u> row, and its 8th number. $15 = (2 \times 8) - 1$.

The 60th move will consequently be the 8th move of the disc two rows below <u>a</u> (Because we divided by 2 *twice*). That is <u>c</u> .

Exercise 12

Describe similarly the 74th, 80th, 1000th move.

The approach

Several strategies were inevitable, provided the Tower Game was played and others probably arose. These will remain as important strategies in learning mathematics and, therefore, in teaching the subject:

1. An entry into a problem was posed. Initially, the problem was to play the game according to the two rules. The entry offered was in the description of the game and the invitation to 'begin playing.'
2. A limited situation was chosen that *every* student could meet – some discs, three pegs, simple rules.
3. The challenge made was also of great potential, for any students who wished could proceed a long way into the playing of the game.
4. Initiative, self - expression and a student's own contribution were capitalized upon from the start.
5. From the beginning everyone was encouraged to discuss honestly what had happened, listening to others if possible, hearing of the practical difficulties of some improvisations, clarifying exactly what *were* the rules and what not, accepting from other players perhaps what the names of the discs would be for identification purposes.

    (*Question*: Shall we talk of disc 1 having the first move? or disc a and move 1? or rod white or rod w and move 1 – if colored rods are used.)
6. If the game of ten discs seems too challenging how can we simplify it without changing its essence?
We first conquered the process for 3 discs, 2 discs, and 1 disc, although to present originally a Tower Game of this few discs would be condescending to the players. Better present initially the vision of a worthwhile challenge, followed by the beginning of strategies to conquer it.
7. Soon discovering that the players find it difficult to remember what moves they have made 'so far', the strategy of recording arises.
What *is* a good method of recording? What shall we aim at? Would it save energy to use shorthand – invented on the spot perhaps – so long as it does not confuse the players agreeing to it? It does *not* have to be composed of standard symbols, those found perhaps in a school or college math text.
8. A reliable set of data is accumulated by trial and error, now that the entry phase is passed. Players check each other and agree on the data.
9. Patterns are examined. They may be intuitive at first, only to become objectified as a student, through practice and discussion, becomes able to say clearly what he sees. From a feeling in one's hand and a hunch that 'this disc should go *there*' can come a conviction that 'this disc *must* go there because otherwise it will have to go elsewhere and that would be against the rules!' (or some other argument)
10. Using the recording strategies, the discussions, the trial and error techniques, and any patterns elicited, the challenges are increased in number. If we are certain of the 2 - disc game, can we now at least play properly part of the 3 - disc game, and then the 4 - disc game?
11. The patterns increased and we began to see a network of interrelating relationships. Perhaps no two students saw the same patterns in the same order, with the same timing, the same significance. If they did we would never know it, but by studying each other's discoveries we could at least reach some kinds of agreement or disagreement.

Mathematical Concepts

A: Exponents and Progressions

The formal mathematics in this is not excessive.

The exponent form for writing some numbers – those called 'powers' – is common, important and can be used by children as early as first or second grade.

2^5 is a short substitute symbol for 2 x 2 x 2 x 2 x 2 which is 32, the standard number name.

2^5 therefore is also 2^4 x 2^1, 2^3 x 2^2, 2^2 x 2^3, 2^1 x 2^4, 2^3 x 2^1 x 2^1, 2^2 x 2^2 x 2^1 ... and so on.

Arising from this we can accept generally that,

$$2^n = 2 \times 2 \times 2 \times \ldots 2 \text{ (using } n \text{ terms, each of 2)}$$
$$2^n \times 2^m = (2 \times 2 \times \ldots 2) \times (2 \times 2 \times 2 \times \ldots 2)$$

   n terms   m terms

$$= 2 \times 2 \times \ldots \times 2 \text{ with a total of } n + m \text{ terms}$$
$$= 2^{n+m}$$

Sequences like 2^1, 2^2, 2^3, 2^4, 2^5 ... and so on are called 'geometrical progressions', an inappropriate name if one wonders what it has to do with geometry. Geometrical progressions are sequences of numbers which progress in the pattern that every term is the previous term multiplied by some number always the same for the particular progression. For the sequence 2^1, 2^2, 2^3, 2^4, 2^5 ... every term (after the first) is 2 times the previous term.

In standard names the progression would be 2, 4, 8, 16, 32 ...

Sequences like 1, 3, 5, 7, 9 ... and 2, 6, 10, 14, 18 ... are called 'arithmetical progressions'. Every term, except the first, is obtained by *adding* a number to the previous term.

For 1, 3, 5, 7, 9 ... the number added each time is 2.

2, 6, 10, 14, 18 ... the number added each time is 4.

Chapters on AP's (Arithmetic Progressions) and GP's (Geometric Progressions) are standard in most high school mathematics texts.

B: A General Proof – Mathematical Induction

We discovered that the minimum number of moves to finish the 3 - disc game was 7. This suggests that for t number of discs the number of moves might be 2^t - 1. It turned out that way when we did it for $t = 4$ and $t = 5$ but does it inevitably follow from that small number of examples that the same pattern will always work?

We had better be careful.

Let us assume that for t discs the number is $2^t - 1$. Consider now a tower of $t + 1$ discs, one more. Suppose they are on post L, the extra disc being the largest and of course, at the bottom.

To move the t disc tower from off this

bottom disc, onto post M, $2^t - 1$ moves are, from our assumption, needed. The extra largest disc is then freed and is moved to post R.

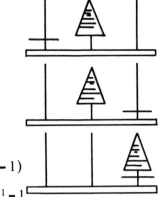

This makes $2^t - 1 + 1$ moves so far.

But if the <u>t</u> disc tower had been moved from R to M it would still take $2^t - 1$ moves. To move it *back* from M to R placing it on top of the recently moved largest disc would also, therefore, take $2^t - 1$ moves. That completes the game, for all $t + 1$ discs are now on R.

$$\begin{aligned} \text{Total number of moves} &= (2^t - 1) + 1 + (2^t - 1) \\ &= 2^t + 2^t - 1 \\ &= (2 \times 2^t) - 1 = 2^{t+1} - 1 \end{aligned}$$

So, if $2^n - 1$ expresses the number of moves when $n = t$ it does also for $n = t + 1$.

If $n = 3$, $2^3 - 1 = 7$, which we already discovered.

The expression $2^n - 1$ also applies, therefore, if $n = 3 + 1$ and if $n = 4 + 1, 5 + 1 \ldots$ and so on, *generally*.

We can be certain, therefore, that no matter how many discs are used, the number of moves can *always* be expressed by $2^n - 1$.

This proof is an example of so called **Mathematical Induction.**

3

Mathematical Games
on a Different Planet

The Planets

The story this time may again seem fanciful and unnecessary though it provides a valuable vocabulary and attitude to topics usually considered difficult and valueless. Following, there will be included alternative settings for the same mathematics. If the prime concern is one of teaching children, each teacher will decide the appropriate materials, taking into consideration his/her own expertise with materials, and their availability.

This setting requires the use of colored rods but it needs no prior experience with them. It would, however, be difficult for a reader to go far unless he/she is prepared to manipulate rods as the text is read. Subsequently, the rods can be abandoned and alternatives used.

The only agreement beforehand is as follows:

the color names of the rods here will be: white, red, light - green, pink, yellow, dark - green, black, brown, blue, orange.

The story begins

You and your companions, if any, are asked to imagine yourselves leaving this earth in a space ship. You are beginning a tour of the universe and travel first to a distant solar system, one which has several planets.

You will land on one of these planets to find a civilization much like your own, human beings, houses, schools, and so on. You are met by an inhabitant of the 'pink planet.' He will be guide and interpreter and his role here will be taken by the author.

The party decides to visit a school, a class of which is apparently engaged in some study using colored rods, for each child has some of these on his desk and often refers to them.

The children at times play freely with the rods, make towers, staircase models, square shapes, trains, and so on. The variations in the use of the different lengths and colors seem kaleidoscopic and the array of patterns seems fresh in its total effect, although on closer inspection most patterns and models are constructed by putting rods end - to - end to form a train, putting two trains alongside each other or placing rods on top of other rods.

Some jargon is used by the teacher and the learners. The length of a train consisting of a red rod and a black rod touching end - to - end is called 'red plus black' or 'black plus red' and on paper written 'red plus black' or in some shorter form like 'r + b' or 'r + bk'.

If a red rod is placed by the side of a black and one concentrates on the length which is necessary at the end of the rod to complete the length of the black, the jargon is 'black minus red equals yellow' or 'b - r = y'.

bk	
r	y

Exercise 1

a. Arrange any rods in a train. Read the name of each train using the jargon.

b. Make a true statement in the jargon.
    examples:"yellow is longer than red"
    "yellow minus green equals red"
    "black plus white equals brown"

c. Invent true statements using any rods and any number of rods.

The teacher on the pink planet is heard asking the children, 'How many white rods equal the light - green rod?' and she is referring to the lengths. The children reply 'three' and they write '3' as a corresponding sign. Asked what are other names for 3 they also come up with $2 + 1$, $1 + 2$, $1 + 1 + 1$. "Yellow minus yellow" gives 'zero' or '0'.

Exercise 2

Write other names for 3 using only 0, 1 and 2 and $+, -$ signs.

It is not long before you recognize that the children can produce a great deal of arithmetic even though, let us assume for the moment, they are only using the numbers whose standard names are represented by 0, 1, 2, and 3.

Inequalities are easy, using the rods to help, though rods are soon abandoned by each child as soon as he becomes aware of some pattern which is more helpful.

For example, once one writes with conviction: $3 > 1$ or $3 \rightarrow 1$ (either the $>$ or the $\rightarrow$ being read 'is bigger than'), then one can also write,

$$3 + 1 > 1 + 1$$
$$3 > 0$$
$$3 + 1 + 1 > 1 + 1 + 1$$
$$3 + 2 > 1 + 2$$
$$3 + 1 + 2 > 1 + 1 + 2 \ldots \text{and so on.}$$

Exercise 3

Write many other inequalities still using only 0, 1, 2, 3, (repeating each as often as one wishes).

Exercise 3 + 1

Sums are expressed (as on Earth):
$1 + 2 + 1 + 1 + 2 + 1 + 2 + 2 + 2$ (a *sum* using 1's and 2's only)
Write other sums; long ones, short ones.

Exercise 3 + 2

Differences, expressions using minus signs, look like these:
$3 - 2$, $2 - 1$, $3 - 2 - 1$
Write expressions using only minus signs; also a mix of $+$ and $-$ signs.

Note that the questions, '*What* is $2 + 3$?' or 'What is the *answer* to $1 + 1 + 1$?' or 'What is the total of 1, 2 and 3' have not been asked!

All that has happened so far is that the children are seen, in our story, to make trains and differences with the rods, compare their lengths, talk about all this and write corresponding expressions, perhaps in the jargon of the color names but also using 1, 2, 3 as number names and $>$, $+$ and $-$ as operational signs.

For convenience, we shall adopt the pink planet teacher's use of the phrase 'standard name' with the corresponding 'non - standard name.' 3 is the standard

name of 1 + 2; 1 + 2 is one of the non - standard names of 3.
1 + 1 + 1,   3 + 1 - 1,   3 + 2 - 2 are others.

Exercise 3 + 3

Write non - standard names for 2 using only the limited numeral vocabulary of 0, 1, 2 and 3. (repetitions allowed!)

Exercise 3 + 3 + 1

Write other non - standard names for 3 + 2,   2 + 1 + 3,   3 + 2 + 1.

Plateau

If the use of a limited vocabulary is new to the reader he is strongly recommended to study it in this context.

Rather than treat arithmetic as an exercise of rushing to the standard names as the only emphasis needed, the view is taken that it is vital to study what can be done with what one is given. The processing of standard names *is* important, but it can come as an important by - product of the study or rich, creative, individually produced possibilities.

The story continues:

To your surprise perhaps, in the pink planet classroom, you one day hear the teacher tell the children:
"Today I shall tell you the standard name for 'three plus one.' It is ten, written 10."

Exercise 3 + 3 + 2

Write names equivalent to 10, still using only 0, 1, 2, 3 as permitted number names, although there is an endless supply of each.

Exercise 3 + 3 + 3

Write sentences which are true on the pink planet. Some of them may be held as true on Earth, but others may be judged as untrue or even 'wrong'.

Multiplication can be introduced. 'Two times three', written 2 x 3 has equivalent names.

Beginning with 2 x 3 the 3 may be changed, for example, to 2 + 1. Appropriate substitutes may also be made for the 2.

2 x (2 + 1)   2 x (1 + 2)   (3 - 1) x 3   (3 - 1) x (2 + 1)... are all names for 2 x 3.

Using the new numeral 10 we may develop other equivalents of 2 x 3:
2 x (10 - 1)   (3 - 1) x (10 - 1)   (!0 - 1 - 1) x (10 - 1)

Returning to the rods 2 x 3 can be represented by two light - green rods in a train. So 2 x 3 = 3 + 3.

3 x 2 can be shown as three red rods. Therefore, 3 x 2 = 2 + 2 + 2 and 2 x 3 = 3 x 2. Piles of counters can also be used to show this.

Exercise 3 + 3 + 3 + 1

Given that the standard name for 10 + 1 is 11, 'eleven'; for 10 + 2 is 12 , 'twelve'; for 10 + 3 is 13, 'thirteen', write many other sentences using number names that are used on the pink planet.

Generally, the process of obtaining a standard name equivalent to any given expression involves finding the corresponding equivalent name using as many tens as possible.

Using rods, take as an example 3 x 3. A train of three light - green rods gives a model. Measure this with as many tens (that is 'pink rods') as possible. In this case two of them are needed and an extra white completes the length. So 3 x 3 = 2 x 10 + 1 and the conventional standard name for this is 21.

Exercise 3 + 3 + 3 + 2

What is the standard name for the number of this Exercise?

Exercise 30

Given that the standard name for 10 x 10 is 100, ("one hundred") create a large number of expressions of your own invention and process their standard names. If you cannot go as far as the standard names in some cases, be satisfied to stop at shorter but still non - standards.

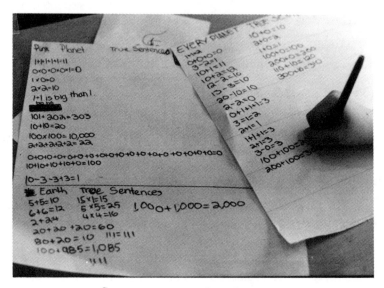

Sentences true on various planets

Exercise 31

Begin with any standard names you wish and produce many non - standards for each. If this can only be done gradually, be satisfied with that.

Plateau

Let us see where we are. So far we have used as written number words only 0, 1, 2, 3 and others which we 'spelt' with two or more of these together. 10 was the first. Then came 11, 12, 13, 20, 21, 22, 23, 30, etc. and now 100.

Their spoken names on the pink planet are exactly the same as on Earth, the same pronunciation. 10 is called 'ten' although clearly its meaning has changed. Instead of denoting the number of dots in this set it denotes the number of elements in this:

If any ambiguity existed at first it was because the visitors had to surrender themselves to some peculiarities of another language, though much of the language seemed as familiar as their own. This is not an unusual experience for any traveller, for on Earth there are many words in mathematics each with different meanings: 'add, set' or in non - mathematical language, for example, 'chair, table, home'.

The challenge then was to compose substitute names for these numbers using +, -, x, ÷ and any other operations known to the reader (such as squaring or exponents). Alternatively, true or false sentences could be written using any of the number names discovered. In all of these the only basic signs available were 0, 1, 2, 3 with the meanings attached to them.

If the reader needs help he can resort to rods, counters or fingers. The color - name of the pink rods identifies a direct connection between 'the length of the pink measured with the length of a white as unit', and the particular planet where that is the convention used by the local community.

There are very many avenues to follow once an entry has been made to the numbering system on the pink planet and when any surprises have been accepted by visitors who, perhaps, were initially put off or puzzled by the different language. One can continue to compose huge numbers of number names and number sentences, as previously mentioned. The only extra information needed is that after 100 comes 101, 102, 103, 110 . . . and so on . . . 120, 130, 200, 210 . . . up to 333. Add 1 to 333 and we get 1000 ('one thousand') which is also seen to be 10 x 100.

Here we shall suggest an exercise in multiplication.

Exercise 32

2 x 3 is called a 'product' or 'product name'. Its standard name is 12, for
$$2 \text{ x } 3 = 3 + 3 = 10 + 2 = 12$$
but that is only one way of seeing it. Using rods, 2 x 3 is represented by two light - green rods whose total length is also 'pink plus red' and that gives 10 + 2 or 12.

Create a table in which the headings are 0, 1, 2, 3, 10 and the entries are the standard names of the products.

When the multiplication table is complete it can be studied. Are there patterns which one can spot?

X	0	1	2	3	10
0	.	.	.	.	.
1	.	.	.	.	.
2	.	.	.	12	.
3	.	.	.	.	.
10	.	.	.	.	.

It is not possible, of course, to anticipate in what order a particular student sees the possible patterns of a table. There are many of them, some more obvious to individuals than others. Without any implication, therefore, that a reader will necessarily choose the printed order of the following questions, we may ask:

a) Are there rows identical to columns?

b) Could one have continued the rows to the right and/or the columns down the page?

c) What are the product forms of the entries in the main diagonal from 0 to 100?

d) Do the numbers of column 2 increase successively? By what number? What of column 3? column 10?

e) Consider any rectangle of numbers in the table like these examples:

X	0	1	2	3	10
0	0	0	0	0	0
1	0	1	2	3	10
2	0	2	10	12	20
3	0	3	12	21	30
10	0	10	20	30	100

X	0	1	2	3	10
0	0	0	0	0	0
1	0	1	2	3	10
2	0	2	10	12	20
3	0	3	12	21	30
4	0	10	20	30	100

Using numbers at opposite corners of each rectangle (or *square*, since a square *is* a rectangle) – is it true that their products in pairs are equivalent?
Does 1 x 12 = 2 x 3?   Does 1 x 21 = 3 x 3?

Exercise 33

Check whether this is true for *any* rectangle of the table shown, or any larger table with headings past 10.

f) Consider the following pattern in the 'pink planet' multiplication table.

In row 1 there are the entries 2, 3. In row 2 the entries in the columns corresponding to the 2 and 3 just mentioned are 10 and 12.

Suppose it is assumed that 'the fraction two - thirds, also written (2,3), is equivalent to the fraction (10,12)', what other fractions extracted from the table are also likely to be equivalent? Could we not look in every row for the entry vertically below the 2 and the entry in the same row vertically below the 3?

This gives (2,3)
$\qquad\qquad$ (10,12)
$\qquad\qquad$ (12,21)
$\qquad\qquad$ (20,30)... and so on, each of them equivalent to each other and called 'two - thirds'.

Exercise 100

By extending the table, extract many equivalents of 'three - elevenths' and its 'reciprocal', eleven *threeths* or eleven thirds.

Playing this game does it seem that (0,0) will be an equivalent of (2,3); also an equivalent of (1,3); of (10,3), of every possible pair?

Normally, however, we want different meanings to (1,3) and (2,3), for we use them as the basis for calling lengths 'one - third' and 'two - thirds'. These fractions are considered different. The pairs (0,0) if accepted here as equivalent lead us into the unwanted position of having to accept (1,3) = (2,3).

(0,0) is, therefore, not considered as having a meaning as a fraction. It is 'not defined'.

The approach

Once again we can check to see which of our former questions apply to what has actually been done.

All that needs adding to the questions of Chapter 1 concerns the possible confusion of calling 1 + 3 'ten', or writing 1 + 3 = 10 with considerable The items listed at the end of chapter 2 can be discussed anew:

1. The entry into the activity was via a background story, using the colored rods for help in manipulation and discussion.
2. The limited situation here amounted to a restriction upon the legal marks which could be made on paper, with their corresponding spoken names. There were 0, 1, 2, 3 on the 'pink planet' with +, -, x, ÷ and = signs. Such a restriction is not new. On earth, conventionally, we limit our construction of number names to the marks 0, 1, 2, 3, 4, 5, 6, 7, 8 and 9, repeated in certain ways to creative infinities of what we want!
3. Even though we only have 0, 1, 2, 3 to use, anyone can produce an unlimited number of other number names and an endless accumulation of true or false number sentences. When 10 is introduced as the standard name for 1 + 1 + 1 + 1 there are more infinities of possibilities. And similarly for 11, 12, 13, 30, 100, 1000, etc.
4. In book form, an author can hardly capitalize on what a reader

invents. But the reader can continually challenge himself by asking 'If I have this, what else can I get?' By so doing, every reader may be persuaded to produce 'a lot from a little'.

5. If a second reader of the text can be found, a discussion can follow.

6. There is no question in the Planet Game of producing *all* the names. For that would be *all* of arithmetic in one base, with significant implications to other but related mathematics!

7. Recording comes naturally when one wishes to overview at a later time what one has developed, or to use the written matter in itself as a trigger to even more development.

8. When one surrenders to the 'pink planet' language progress is rapid. One finds no difficulty in accepting statements either in the context of the 'pink planet' or that of the 'Earth' (or 'orange planet').

9. Patterns certainly emerge from the beginning of the work on the 'pink planet':

    e.g., if $1 + 3 = 10$ then $3 + 1 = 10$, $10 = 1 + 3$, $10 = 3 + 1$, $10 - 3 = 1$ . . . and so on.

    if $2 \times 3 = 12$ then $12 \div 3 = 2, \frac{1}{2} \times 12 = 3$ . . . and so on.

    $(1,3) = (2,12) = (3,21) = $ . . . and so on.

10. Once one can operate on the 'pink planet' one can venture onto any other planet, the meaning of ten (10) always being associated with the number of white rods equal in length to the rod which bears the color name of the planet.

    The only reason why normally it seems difficult to go from Earth language to Pink Planet language is because we may be meeting the first departure from what we believed to be of universal significance. When liberated, it is much easier to go from 'planet to planet' than it was initially. This is true in this context as it was for man's first passage from this planet to the moon. It is only technical reasons which prevent further trips, not the knowledge of whether it can or should be tried!

11. There are very many patterns omitted in this brief treatment. Some of them will follow in other chapters. Every reader will find others not mentioned.

The mathematics

The mathematics of this is, of course, called 'change of base', 'different bases' or 'bases of numeration'. Its importance is as follows:

a) it provides insight into the fact that numbers and their operations act similarly, whatever the base. It is the *language* used for the numbers which changes, not the numbers themselves or their inter - relation-ships.

b) The coding of the 'red planet', conventionally called Base 2, is used in computer work. So is Base 8.

c) The historical development of number systems in various countries of this planet provides evidence of different bases being used in different countries. For example, the use of 'quatre vingt' in French suggests a remnant of base 'twenty'.

Normally the symbolism used in traditional mathematics books is like this:

$$524_{seven} \; = \; \left[\; (5 \times 7^2) + (2 \times 7) + 4 \; \right]_{ten} = 263_{ten}$$

Probably two aspects are responsible for the little work accomplished in schools in 'different bases'. One is the above illustrated process of translation which is emphasized from the outset. With spoken languages it is being increasingly accepted that translation from one language to a second is not conducive to a student's learning. Only when he is very competent in both will he be able to go from one to the other with ease. Before that it is better to learn each by being immersed in each.

The validity of this last statement can be believed by noting that all young children, the world over, master their own first tongue – that usually of their parents – regardless of the particular language. Yet if at a later age we learn a second language by the traditional method of translation, we find it very difficult to avoid interference and our minds seem less likely to be fluid and creative.

In the approach we have taken translation, therefore, is not mentioned. The meaning comes directly from the use of the materials and the coding names which come from the teacher. He does not say, as so many mathematics teachers do, "There is no 5 in Base 5," which seems to be a contradiction in terms! When 10 is 4 + 1 there is no standard name for this, other than 'ten'. Its referent – to a rod length or to a set of fingers – indicates that *this* 'ten' is not the same as 'ten' in other bases ('planets').

The second difficulty is in the claim that 10 in any base other than the common base should be read 'one - oh' or 'one - zero'. The intention is admirable: to clarify that 10 in one base is not in all ways the 10 in another. But ambiguity is not avoided if one progresses very far. 100 will have to be called 'one - zero - zero', and 1000000 'one - zero - zero - zero - zero - zero - zero'. Why not simply call this '1 million' so long as *everyone knows the context?*

The context quickly becomes clear with the planet story. If rods are not used, some other legend needs to be signalled to all. Pebbles, counters or dots on the blackboard suffice to act as a referent for any student needing a check. ⊙ - 10. Any fictitious name can be associated with this referent. Chosen by the students themselves: "In Timbuctoo ⊙ is called ten." "In Wootton - under - Edge ⊙ is called 10."

It is probably for such reasons why the study of numbers and operations in many bases has faded somewhat from official curriculum requirements. Approached traditionally it seems too difficult, too confusing. The alternative begun here has quite different effects once the initial shock of discovering that familiar names may have changed their meaning has been passed. Progress can then be made rapidly and joyfully and the awareness comes that one can operate almost as easily in any base, *one* of which happens to be that which associates 10 with the fingers of our hands.

Moreover, much mathematics in the sense of underlying patterns is true equally in any base. Only some of the names change. Structure does not. There is no need to memorize such groups of facts as multiplication tables, except in the base that one's local society uses constantly in commerce and trade. One has, through experience of the structures at first hand, the confidence of being able to reproduce any required development in *any* base fairly quickly.

This is, therefore, a rich opportunity to master the awareness that in mathematics it is the *trip itself* that matters. Knowledge and the reproduction of *some* of the incidents on the trip will probably be retained. With the ability to work within the structure one can always, if need arises later, begin from what one *does* remember, and recreate portions of the trip by operating as one has done so many times previously.

The Tower Game again

Now that another enumeration system has been introduced turn back to the end of the Tower Game in chapter 2.

In binary notation (red planet language) the successive moves of the disc appear thus:

Disc a 1, 11, 101, 111, 1001, 1011, 1101, . . .
     b 10, 110, 1010, 1110, 10010, 10110, 11010, . . .
     c 100, 1100, 10100, 11100, 100100, 101100, 110100, . . . and
so on.

Note that the only difference between rows is the number of 0's at the end of every number name.

Those in row a have no zeroes at the end.
        b each have 0 at the end.
        c each have 00 . . . and so on.

Exercise

What is the 1101th move?
It is odd, so disc a will be on the move.

As previously it will be the $\dfrac{1101+1}{10} = \dfrac{1110}{10} = 111^{th}$ move of a .

(The 111th move on the red planet is, of course, called the 'seventh move' on the orange planet.)

All this may appeal to some people as simpler than in the standard base!

Exercise

What is the 111000th move?

The presence of 000 at the end in the move of the above Exercise indicates a move of disc d . The rest of the name, 111, indicates that it is the fourth move (if we retain the common language) or the 100th (if we remain pure to the base) move of that disc.

Summary of Chapter 1 - 3

We set out with the purpose of involving every reader, as far as one can do using only the written word, in some activities probably not previously met, which turn out to produce a substantial amount of enjoyable mathematics.

Readers were asked, as they went along, to study what they did, what they felt, what successes they had and where they seemed to stumble or become confused. It is claimed that almost inevitably all readers followed certain patterns of development – of *learning* – which can be present in not only all mathematics lessons but in lessons of all kinds.

An entry point is first needed to get one going on the task but in the early stage the issue, problem or study has to be clarified. There is hardly ever a circumstance such that a learner knows immediately and clearly what has to be done. There is a groping stage and any suggestion that a lack of clarity means that the learner has not 'paid sufficient attention' is, in one sense, unfair though, in another sense, it is true. But 'sufficient attention' may not be possible for some time. Decisions may be needed as to what is relevant and what not, so some groping is inescapable. The time, details and intensity of this first stage will vary with the occasion, the nature of the problem and the unique experiences of every individual learner.

There then follows – if anything – a period when each learner begins to grasp what are the essences of the problem. He enters more deeply, organizing things according to his own perception of what is present and perhaps makes recordings. Such development may not be smooth or continuous. There may be pauses, plateaux for rest or recapitulation and assessment. They may last for minutes or be taken overnight or even longer. By this time parts of the task will have been mastered, at least perhaps the problem itself will have become clear, so during subsequent plateaux there may come awarenesses needed to be sought for completely mastering all that is challenging.

Finally, on condition again that movement forward is still occurring, the mastery of this and that part of the problem synthesizes to a mastery of the whole. Other recapitulations may follow of the entire study, fresh subproblems may be put by the learner himself and solved and there is present a sense of accomplishment – maybe even one that after all the striving it seems odd that it took so long to see the necessary clues, strategies or understanding of what was needed! The problem has been mastered and fresh challenges are sought.

The three stages of learning briefly described are not clearly isolated, for in the same task a learner may be working in the 'grasping' stage in one part of the problem while groping in another. Prior experiences may help or hinder, any materials involved or brought into the problem may provide clues or they may obscure. Discussion with other people similarly engaged may trigger one's own thinking, helpfully or unhelpfully.

The passage from initial confrontation and a decision to engage, through to complete mastery may not always seem clear but all of us know about them whether it be in our conquest of walking, riding our first bicycle, learning our mother tongue at the age of 2 and 3 or in the beginnings of arithmetic.

Preview of Following Chapters 4 - 17

We have briefly described learning from the learner's point of view. Much more needs to be said but not in this book. The emphasis here will be upon the characteristics of mathematics needed to accommodate every individual's learning complexities. For powerful patterns of the subject matter must fit the patterns of the learners.

Instead of presenting mathematics in the traditional mode the strategies will be by way of invitation to the learners to become involved in various activities – just as was done in the first three chapters. Henceforth, however, the topics considered will be those familiar and common to school arithmetic and mathematics. As far as a book is able to do so – without being able to respond to a reader – some of the possible mathematical patterns will be stressed, readers being encouraged to express their preferential patterns by making up examples of their own.

Traditionally, a teacher presents a set of exercises for students to tackle. This finished, everyone seems to have the impression that there is nothing more to be done. The teachers and the authors of texts have, of course, the best intentions of providing relevant examples to what is being studied, but unless greater demands than this are made how can we be sure the students understand other than superficially? By having learners continually create their own mathematical examples they are more likely to understand the power they have to attain fluency.

Much of the approach we advocate is characterized by the words of the title ' . . . and So On', because:

a) it is a phrase common to mathematical texts, denoting a continuing pattern, with a 'glimpse of infinity';

b) the practice of proceeding further than the explicit examples given, helps to induce in learners more responsibility of dependence on themselves. It tends to place the authority of what they do on their own invention, experience and imagination, rather than on some outside authority of teacher or answer book. The latter have too frequently stressed that it is the result, not the process, which matters most;

c) it encourages an 'algebraic awareness', because the emphasis is placed on generalized understandings, not merely on answers to particular problems. This is a more powerful way of working with mathematics, dealing with rich ideas and connections between what otherwise may seem to be disconnected topics;

d) the implication is that every student can always continue with what he has begun, even though later it may have to be revised to fit with his own changing hypotheses or those of others. It also provides greater challenges for those who wish to delve deeper into a study.

As we shall see, all this is in accord with what is later needed in more formal algebraic studies.

Traditional mathematics teaching can be said to be of a convergent nature. Teachers have set the problems, students have gathered together the data provided, operated on them and narrowed it all down to an 'answer', hopefully considered to be correct. Attainment of that end has been the dominant objective. We may, however, ask why the emphasis on right answers has not produced for more adults a pleasurable sense of high competence in mathematics or seen their efforts in the subject as a worthwhile investment of time and energy. May it be that big changes are needed in the understandings and practices of the teaching *and* the learning?

One change we propose is to work for divergency. One in which, for example, the answer is sometimes given, the challenge being to invent the problem! This not only makes for better mathematics; it is also a more sensible way to accommodate the complexity of individual differences of students.

Algebraic Emphasis in Children's Learning

An algebraic attitude to the learning of arithmetic is important because all children by the age of 5 have been using such a mode in their learning, especially noticeable in their learning of language. We do not mean the kind of algebra seen in the secondary school textbooks. The notion is developed in chapter 4 but an analogy here may be useful.

For instance, an adult may see a small child using only one chair for experimentation while he, the child, studies for the first time 'how to sit on a chair'. However, even if the young learner literally has only ever encountered *one* chair he in fact learns to sit in *any* chair he comes to in the future. *To the learner* the one chair represents all chairs. He generalizes (correctly in this example) because the ability to sit on a chair lies in the use of his self – for balance, muscle tone, muscle movement, relaxation, stress, skin sensation, etc. – and when the person has mastered 'sit - ability' that mastery can be applied to any chair which is deemed to be like the first experimented on. Of course, there may well be extra value in meeting other chairs, to get a wider appreciation of differences with the sameness of chairs. The child may also have to amend some of his hypotheses made early on with the first chair, but amid all this there are powerful generalizations latent in what has been experienced.

Such experience with generalizations can later be recognized in the mathematics classroom even if with some topics the teachers know that some students' hypotheses of generalization are not going to stand up when pursued further. Teaching strategies, consequently, can ensure that a wider experience becomes available to the students instead of those initial hypotheses being blocked by someone continually saying 'wrong'.

What Mathematics Is

To convey through various mathematical topics some application of algebraic awareness which harmonize with the way children learn, we need to consider a little more closely the nature of mathematics.

Mathematics is concerned with patterns, with relationships, with relationships of relationships, with operations, expressing it all in codified sounds, ink marks and drawings. We shall for the purpose of discussion sometimes appear to separate the mathematics from the language used to convey it, although both aspects are intimately bound together just as are meanings and words, phrases and sentences of the common 'non - mathematical' languages. Nevertheless, we know that the words in themselves have no intrinsic meaning. The word 'yellow', for example, is not yellow; it is only *associated* with the color. We shall, therefore, meet in this book continual references to number *names* in distinction from the numbers themselves. They are the virtual experiences 'behind' the names which *can* be seen and heard.

Mathematics includes many topics: different systems of numbers, collections of numbers, systems of points, planes, lines . . . and so on. A great variety exists, with more specializations being created by professional mathematicians such that none of them can keep abreast of what all the others do. Arithmetic is one of the topics first introduced into the public school system a century ago because its results were needed for the great increase of the number of jobs during the industrial revolution period.

Arithmetic, however, has not, in the past, been treated very mathematical-ly but seen as an accumulation of facts. We are now suggesting that only a command of the processes by which the arithmetic is understood can possibly ensure that it be a worthwhile task for everyone. Such processes include the patterns of the learners, their individualities, their samenesses and their differences. They also include the patterns of the mathematicians but which, when presented later will not be contradicted, but are introduced in ways and means more appropriate for children.

Numbers and Number Names

Not every mention of a number name implies that arithmetic is being studied. In the tables of "My Three Sons", for instance, some number names were first used for descriptive labels only. The names 1, 2, 3, 4, 5, 6 are certainly arithmetic forms, but for the names of the various 'change - overs' no properties of the number names are important. The headings for the tables could just as well have been letters of the alphabet or other labels.

Arithmetic came into chapter 1 with a question like, "How many combina-tions are there for two change - overs?" since what was wanted in that context was a name which arises from counting and from the realization that any number label will *not* do, because the number of change - overs is what it is and not the number of some other set. The standard name for the reply was in this example, of course, 'thirty - six', though it might have been given by a non - standard name such as 'six sixes'.

In chapter 3 both arithmetic and non - arithmetic mathematics were present. The fact that $10, 1 + 3, 1 + 2 + 1$ ... and so on, are some of the names for 'ten' on the pink planet, makes this kind of exercise arithmetic because the emphasis now is on the production of alternative names, some of them standard, others not. On the other hand an awareness that 'sums can be reversed' or that 'the addition of a number followed by its subtraction is equivalent to the addition of zero' – these statements are more mathematical, more algebraic, because they are generalizations and more powerful than merely knowing $1 + 3 = 10$.

To work with an algebraic approach in arithmetic the symbols do not have to be those traditionally reserved for Algebra. $a + b = b + a; a + x - x = a$ would be formal and traditional ways of expressing the two principles quoted just above. An interesting illustration of this might have occurred during a reader's sojourn on the 'pink planet'.

Back on the 'pink planet'

What would the sentence $4 + 1 = 10$ mean on the pink planet?

A first reaction might be that such a sentence would never occur; 'there is no 4 on the pink planet'. This is a misunderstanding, however, for there is no way of stopping anyone making the mark 4! The fact is that 4 is not used as a number name on that planet, not that it is not used. It could be used as a place holder or *as a variable,* or 'unknown number' like the alternatives $\square + 1 = 10.\underline{\quad} + 1 = 10, x + 1 = 10$. Each of these is an open sentence because it is not possible to assert the truthfulness or falseness of such incomplete sentences. It depends on what can be substituted for the 'box', the space, the letter or the 4.

$4 + 1 = 10$, therefore, is an equation and it can be solved:

$$4 + 1 = 10$$
$$4 = 10 - 1$$
$$\therefore 4 = 3$$

This indicates the value for mathematical growth of operating in symbols other than the conventional, to realize that structures underlie the particular outward codifications and to be able to change easily to other forms.

Summary

In this book the topics of elementary arithmetic/mathematics will be studied but the treatment will be algebraic in non - formal ways. Certain concrete aids and manipulative materials will be discussed because they are not only important for learners as referents but those chosen are richly open to algebraic insights. It is not implied that such aids are for primary children only or for those students who 'do not get it' in traditional paper and pencil form. They are essential models from which come directly - seen relationships which can then be talked and written about. Their use is not confined to any special kind of learner or level of learning. They are relevant to the mathematics of all school grades.

The first elementary topic to be considered is that of the importance of language in arithmetic computation.

Tables on various planets

4

Computation as a Language Art

We shall now develop assertions that a young child's learning of his first language provides evidence of his potential for his mathematical growth. It is because it is during his mastery of the common language that he used structure, patterns, abstractions, transformations, signs and symbols spoken at first and only sometime after that put also in written form. These and many other operations and viewpoints are also at the core of the mathematics. In particular the essence of arithmetic is one of language manipulation the likeness of which the learner has previously met and surmounted. There is every reason therefore why this rich experience should be applied when the topics and meanings of the language now refer to numbers and what can be done with them.

First, arithmetic is a language which is so like our common, non-arithmetic language that it is odd that the similarities have hardly been recognized in school work.

Second, all young children (with very few exceptions) have great success in learning their mother tongue whether it be English, French, or any other, and they do this without formal instruction at the ages of 2, 3, 4 and 5 – before they enter Kindergarten.

Third, the essence of the computational aspect of arithmetic is the substitution of words for other words. Children, by their conquest of language, show they have that skill.

Fourth, the substitutions in the arithmetic language are easier than those called for with the common language because in arithmetic there are so many patterns and clues built in to the basic structures of the vocabulary.

Fifth, the key to teaching arithmetic well is to utilize everyone's powers of language and pattern making. This, in arithmetic, is controlled by the mathematical meanings which underlie the signs and symbols, effected by the use of a variety of activities, materials and strategies.

Arithmetic as a Language

Arithmetic is an activity written and talked about. Its language of course carries meanings, some of which are agreed upon approximately by people using the language. But not always, for many words carry ambiguous meanings and are one of the great causes of arguments. The participants may be using the same words, phrases and sentences but the meanings are different.

Arithmetic *is* a language and when we *do* arithmetic we are engaged in a language art. The art of arithmetic is *one* of the arts of language. Let us think about this for a while, since most of us were probably conditioned by schedules in school to believe that, when the reading lesson ended and arithmetic began, there was no more reading! It may not have occurred to us that arithmetic books are *read*; they are *written* by someone with the intention of being read. Maybe even talked about! In fact, arithmetic books are just as much reading books as those with stories in the previous lessons. The stories are different only because they deal with aspects of people's living not usually associated with novels, fables, myths or other things we all do.

Another difference is that number stories are usually arranged inschool textbooks with pauses in between, purposely arranged by the author. Questions are inserted about what has gone on so far. They are called 'Exercises'. Not only that, questions are asked about characters sometimes not always introduced in the immediate episode – or so it frequently seems to the readers! Such questions in a non - arithmetic storybook would not only seem odd, but even unfair, especially if the readers felt strongly they *ought* to know about the strangers as a result of being told only about the main characters!

Let us list some of the ways in which an arithmetic book is like a reading book from the viewpoint of language.

1. They are both written (or printed) and intended to be read.

A long rod train

2. The writings imply meanings.
3. The reading book concentrates on stories about people, events and so on; the arithmetic book upon stories of and the characteristics of such things as numbers and lines and what can be done with them.

4. The A - book (Arithmetic book) contains many number names, phrases and sentences because they are its main characters. It also contains common words, phrases and sentences.

For example, 32, 7 , 964 are names of numbers, number - names. So are 'thirty - two, seven, nine hundred sixty four', expressed in the common *written* language but rarely used because they are so cumbersome. However, both '32' and 'thirty - two' are spoken and read exactly alike. There are always the two forms of number words, but for common words there are not normally two forms. (But we *do* have 'e.g.' and 'for example', 'Mr.' and 'mister' '. . .' for 'and so on', . . .)

$30 + 2$ is also a number name. Its *standard* form is 32 but there are not only many more other *non-standard* forms – there is an endless number of them! $31 + 1$, $29 + 3$, $28 + 4$, 4×8, $321 - 289$ . . . and so on.

$28 + 4 = 32$ is a number *sentence* . It can be interpreted thus: "$28 + 4$ is a number name. It happens to be read here in English although this written name is recognized internationally. Now, there happens to be in English and in the international code, other names for the same number, *one* of which is the so - called standard. It is 'thirty - two' written 32. To write $28 + 4 = 32$, therefore, or say "twenty - eight plus four equals (or 'is equivalent to') thirty - two', is to convey that 32 is the appropriate substitute standard name for $28 + 4$.' "

But we can also say that, corresponding to 32 – the noun for a certain number – there exists the corresponding compound - noun: $28 + 4$, and this can be used on occasions as a substitute for the standard. There are also, as noted, many other substitutes for 32 and for $28 + 4$.

In this substitution activity we are involved in words, sentences and the meanings we attach to them – just as with the common 'non - number story' language.

5. There are correct spellings in language writing and incorrect spellings. So, too, are there in arithmetic. A child who writes $7 + 8 = 51$ and reads it 'seven plus eight is fifteen' spells the numeral for 'fifteen' incorrectly. Its correct form is 15.

This is a common misspelling in early grades, not surprisingly, for one might expect the 5 to be written *before* the 1 since 'fif' is *spoken* before the 'teen' in 'fifteen'. Moreover the words 'fifteen' and 'fiftyone' almost rhyme, thereby blurring clear awareness of the difference.

Misspellings in A - language are more difficult to notice because they are easily accepted as other words, correctly spelt! In the example quoted above a student *may* mean $7 + 8$ is 'fifty - one'.

6. There are true, false and open sentences in both A and common (C)

languages.

	A - language	C - language
True	$8 + 5 = 13$	All cats were once kittens.
False	$5 \times 9 = 27$	New York City is in Canada.
Open	$25 - 7 = \square$	My name is □□□□
Open	$25 - 7 = \text{---}$	My name is ————.
Open	$25 - 7 = ?$	My name is ?
Open	$25 - 7 =$	My name is

The 'open' sentences are not complete. It is not possible to say whether they are true or false. One of the most common activities in the traditional teaching of arithmetic is to have pupils complete open sentences, although this is *not* so emphasized in common or written speech.

7. Punctuation is important in both languages: The child sitting on the chair wearing a bright shirt . . . (pretty chair!); $6 \times 5 + 4 = 54$ (or 34?)

8. Efficient reading of both languages includes reading words, or parts of words, *not written* :
Dr. Smith, South Dr.: "Doctor Smith, South *Drive* '
172: "one *hundred* seven *ty* two"

9. Signs and symbols are not always read in identical ways. The contexts give the clue to the appropriate reading.
'mean' is read differently in 'I *mean* to say . . .'
and 'my *mean*derings . . .'
X is read differently in 6×7 for 'six *times* seven'
and 6×7 for 'six *multiplied by* seven.'

10. Signs and symbols read identically sometimes carry different meanings.
"I write on the table."   Is 'table' an article of furniture or some writing on paper?

$$6 \times 4 \qquad \frac{6}{7} \times \frac{4}{9}$$

Does x mean 'repeated addition' or something else?

11. Legible writing is important though often with the common language the general meaning will hardly be affected by the illegibility of a word. In A - language a 6 written like a 0 *may* cause a significant difference of meaning.

12. Words can be constructed in both languages from combining the same signs in different orders. With a , p , t , the 'words' apt, atp, tap,

tpa, pta, pat are possible. Only three of them are accepted as English words.

With 3, 4, 7 the number words 347, 374, 734, 743, 473, 437 are possible. *All* of them are acceptable as names of numbers.

13. In both languages words can be compounded from combining individual words. Twenty and four compounded give twenty - four. Home and coming compounded give 'homecoming'; 2 and 4 compounded give 24.

14. There are, in both languages, nouns, pronouns, adjectives, predicates, punctuation marks. $6 + 7$, 13 are nouns. x can be a pronoun or 'pronumeral'; $>$ and $=$ are verbs.

$6 + 7 = 13$ is a sentence: $6 + 7$ is the noun subject
$= 13$ is the predicate
$=$ is the verb

13, another noun, is the subjective complement of the verb.

Discovering relationships using colored rods

Children Learning Language

Having argued that Arithmetic has so much in common with language, it is appropriate to pause and consider in broad terms how young children do, in fact, learn their *first* language, their mother tongue. More and more evidence is being produced these days of what happens.

Without referring to technical research, however, it is probable that most people are aware that children around the age of 2 or 3 do learn their mother's tongue and that in the following period of two years learn to function in that language to a remarkable extent. Remarkable in that few children later in life will accomplish so much language learning in a comparable time.

Moreover, the learning takes place without formal instruction. It is hard to believe that any mother tells her infant what to say or how to say it. If she tries she finds that, of course, the infant does not understand her language let alone any possibility of its taking heed!

Traditionally, the matter has been settled by saying, 'The baby imitates' but a moment's thought, common sense, and common observation will suggest that imitation cannot occur very often. The 2 year old does not imitate much of what its parent says. It can not copy the movement of her tongue, larynx, vocal cords or diaphragm yet the movements of all these to produce intended sounds have to be learnt. On the contrary, many parents frequently despair at their inability to get their child to imitate! Let them consider, for instance, trying to persuade their sons and daughters to verbalize whatever common expressions of courtesy they would like them to use!

What is more, the young children learn their first language while engaged in many other tasks. If they do concentrate for a while on some aspect of language study it is seldom obvious to onlookers. Even more seldom do the learners announce that that is what they are doing. Over two years the children play, eat, sleep, lie around the house, watch T.V. and lead lives which frequently seem quite separated from that of their parents. Bodily, of course, they may be sitting at the same table, in the family car or in bed. Within themselves, however, the language is being acquired.

Only when the children reach school will any formal language instruction occur. Then it unfortunately may be done in predominantly *silent* circumstances for each individual. Even if there is plenty of reading aloud and verbal discussions in activities like 'Show and Tell', it is seldom that *arithmetic conversation* is stressed.

It may well be asked if indeed arithmetic is a language and if it can be learnt as other languages are, whether it should not be talked about by children far more than is usual. Of course, we need to note that the main subjects in the arithmetic 'stories' are abstractions and it has certainly been believed that this makes it more difficult to understand than the people and events which feature in tales and in the common language. Nevertheless, a five year old child has little difficulty in talking and listening to stories of kindness, beauty, ugliness, and colors – other abstractions.

Other points are worth noting: First, children in the early months or years of using the common language do not tend compulsively always to strive for standard names. 'That man', 'the person with the mailbag' are ways of referring to someone whose standard name is 'the mailman'; 'mother', 'mom', 'mommy', 'ma' may all refer to the lady whose standard name is Mrs. Smith. There is no reason at all, we shall infer, why children learning arithmetic should *compulsively* be taught only to strive for standard names. Asked for an equivalent to '6 + 4' they are quite capable of saying other non - standards, even though the standard '10' can also be answered.

Second, children do not learn language solely by memorizing a collection of phrases, though they do retain phrases. A phrase book when visiting a foreign country is an admirable aid for certain purposes but its use hardly justifies a claim to knowing the language. To do that one needs to function efficiently to create sentences *never before thought of or heard,* by combining in the acceptable structure of the language certain invariant words and perhaps phrases, with variations of order, stress and length which are appropriate to new situations.

Concentric circles? or polygons? or something to do with clocks?

Children *can* do this in the common language and they *do.* It will be claimed here, therefore, that to neglect this power in the development of arithmetic language is not only to deprive the children of what will produce far better than normal arithmetic results, but it explains one of the faults of traditional approaches – the lack of encouraging the students to create their own arithmetic, having substituted for it a dependence on fitting into the phrases of other people printed in textbooks or on worksheets.

Third, there are phrases in language learning which begin with the music of the language, in that when listening to an infant 'talking' one can detect before any words are clear, the distinctive tones, cadences, rhythms of the language he is beginning to learn. Gradually from this set of sounds, over months, some recognizable words are heard and phrases are uttered, even if of only one or two words. While skill grows in the acquisition of the number of phrases and words available to the child, it will still be months or even years, before finally settled pronunciation is decided upon by each individual. Undoubtedly the learner will be affected by what he hears and there will be corrections and some imitations.

But poor pronunciations will persist far more than could reasonably be monitored, let alone corrected by adults around. Yet, despite – or is it because of? – the adults' inability to correct, or to try and correct all the wrong words or mispronunciations, most learners end up by using the accepted, or one of the accepted, conventions of his immediate society. The child's word 'broked' becomes 'broken'. The 'me wants it' is dispensed with for its substitute, 'I want it'.

The main point in this is that the majority of children attain what is accepted by most adults as correct, *without much formal* correction. They do it by becoming aware of the samenesses and differences between the sounds uttered by themselves and those others make. Such awarenesses will occur in respect to particular words or phrases, at different times and in different orders for different children because of the enormous number of variables operated upon by the learners. Such awarenesses seldom occur unless the learning children are subject to a great deal of talking by themselves and by others in their presence.

The powers that the children use for conquering the speaking of their mother tongue are, of course, available also for 'speaking arithmetic'. Speaking it in as many meaningful situations as possible, just as with the common language.

Computations as Substitutions

One of the most important parts of A - language is that which concentrates on computation. The conquest of this comes from being able to develop the skills needed to substitute words for other words – already accomplished in the common language – but where|there now exist special|very helpful rules for creating endless substitutions. Let us examine this further.

In the common language there are not many rules for finding substitutions, though there are some. One has to know, by being told directly or indirectly, for example, that the word 'bunny' is an acceptable substitute in English for 'rabbit'. The use of prefixes on the other hand, can help us to produce a substitute without our being told in particular cases. If we know that 'minibus' is a small bus we could invent 'mini - house' for a small house and while not used much the meaning would probably be clear. And if we don't know the name of Mr. Smith's wife we can always try Mrs. Smith even though, these days, we might be wrong if the lady had not at her marriage adopted her husband's last name!

For computation it is still true that *some* substituions just have to be remembered. If a learner wishes to use these basic words he has to be told what they are together with their associated meanings. After that, however, he is in a position to make up millions of others for himself, as we shall see. For example, 'six times seven' can be substituted for 'forty - two' in appropriate contexts, and vice versa. But no one could possibly know that these two nouns were equivalent, in that one could be substituted for the other – unless either it was invented by oneself or, alternatively, it was accepted by surrendering to what *other* people said.

However, for more complicated examples, we do not have to recall quickly the required standard substitutes. If we did have to remember every possible example either we would have to try and memorize thousands and thousands of standards corresponding to non - standard forms or we would have to remain content with familiarizing ourselves with relatively few. In the latter case we would have no guarantee what to do with the constantly appearing new examples in business, economics, trade or measurement.

Fortunately, only a few basic words have to be remembered. From these, a set of procedures follows to give us the rest. Given 68 x 74 for instance, we learn the pattern that four separate words are first teased out ready for substitution: 'sixty - times - seventy', 'sixty - times - four', 'seventy - times - eight', and 'eight - times - four'. Of course, we do not write them like that but as 60 x 70, 60 x 4, 70 x 8, 8 x 4.

32 is then substituted for 8 x 4, 70 x 8 is related to 7 x 8 for which we substitute 56, 60 x 4 comes from 6 x 4 = 24 and gives 240, 60 x 70 comes from 6 x 7 = 42 and gives 4200.

Finally, the procedure involves the substitution of 5032 for 32 + 560 + 240 + 4200.

This may all seem longwinded for there is so much pressure in schools to get the 'correct answer' quickly. But the correctness of this depends precisely on the awareness that substitutions are involved and on procedures to get valid substitutions. It is, therefore, one of the arts of language and is surely learnt as language is learnt:

> with mistakes corrected by oneself;
> with practice in meaningful situations;
> with new vocabulary or rules embedded within what is familiar;
> by coming up with various possible non - standards if the standard cannot be read immediately;
> and by creating sentences new to the individual.

Attention to such processes must be paid by learners and teacher, therefore, just as much for arithmetic as for other languages. The only differences between the two languages are in the meaning of what is talked about, and that there are two systems of signs for the arithmetic, the numeral version being the one which lends itself especially well to the processes by which all the acceptable substitutions can be made.

The Beginnings of Arithmetic Patterns

All children learn to count before they reach school at age 5, even if they don't always do so 'correctly'. At age 3 they use number - names. 'Three' is naturally a common one at that time and most children also then talk about being 'four' or 'five'. Number names become frequent in their conversations. Even 'large ones' like million and thousand are used somewhat relevantly though not necessarily used as correctly as one day they may be.

A child who first counts may use, for example, "four, three, nine, six" in touching her fingers of one hand. Although "one, two, three, four", is, as we know, the correct English in this context, it is nevertheless interesting to note that the girl has used words which are number words. She has apparently already noted what are some number words and what are not and that is a necessary beginning. Parents should be delighted to hear 'four, three, nine, six' or 'one, two, three, twelve, fourteen' and such like, rather than 'four, three, dog, six' with the intrusion of a non - number word!

Gradually, every child amends this, making adjustments in harmony with the language of the people in the immediate environment, which these days includes the help of T.V. The child begins, in other words, to count more correctly. (Not 'in other words', but in 'other *orders* of the words' already used. Pun intended!)

Note, however, that the children have first to associate the words 'one, two, three . . .' in the conventional order when there are no intrinsic clues to the correct way other than that is what people say. There is no more 'twoness' in the word 'two' or the numeral 2, than there is wood and four legs in the word 'table'. Not even the 'meaning' helps. It is not possible to know that the word 'five' is associated with an important attribute of the set   ⬭   unless one invents it for oneself or accepts from others that that is what *they* use. However, if one *does* associate 'five' with   ⬭   and 'two' with   ⬭   to know that 'five' is regarded as greater than 'two' no further information is needed. The relationship 'greater than' is inherent in the direct experience one has with ·∷   and   ·.   and everybody can have that.

A discussion such as this is important because in the teaching of arithmetic some clarification is necessary between a traditional view of 'telling' and a more modern approach of 'discovery'. In fact, as the above examples show, *both* techniques are inevitable because that is how people learn. Some things have to be told, some things do not. The latter may be directly experienced or inferred from patterns and clues.

Beyond 'nine' and 'ten'

For a while, therefore, a young learner of counting has to grope with new words, each to be associated, he gradually discovers, with particular piles of pebbles or fingers or lengths of colored rods, and with no other – just as the adult learner had to do in the work with different bases (Chapter 3) if that was new to him. The child realizes that if he plays the game of reciting these new words there are correct orders and others have to be abandoned. "One, two, three . . ." is a correct order. So is "ten, nine, eight . . ." "Ten, two, three . . ." is not normally accepted as correct although interesting games can arise in which children may use such orders, checking as they go to see if each name has been used and once only. Games on the fingers sometimes stimulate these.

It *may* cross some children's minds, though this is by no means certain, that there are so many numbers to which names have to be associated, that the sheer retention of the vocabulary yet to come seems to be a prodigious task. There is little evidence, however, that children are ever in awe of a vast vocabulary of words in the common language which we as adults know exist. The children do not view the contents of a dictionary to be a great language obstacle, for they are masterful, as we have said, in the games of substitution and repetition and are content in the powers they have with the vocabulary they also have. As adults, however, we can notice how fortunate it is that there is *not* a vast vocabulary of arithmetic words ahead of the basic few, for most words to come are compound nouns constructed by putting together some of the words already possessed.

When one has learnt the words 'one', 'two' . . . and so on, up to 'ten', with the correct order and applications perhaps to concrete counting situations, the task of counting becomes easier and easier, not harder as has been implied by arithmetic curricula the world over. Those seem to imply that children should in their early studies of numbers only concentrate on the 'first ten' for a year or so, before they are allowed to meet (in school anyway) eleven, twelve ... twenty, and certainly before they are encouraged to deal with hundreds.

Yet as far as the words are concerned – the *names* of the numbers – the process by which new names are produced is merely a matter of forming certain compound words, every one of them different from every other, yet constructed from combinations of the basic names of the first few.

With those basic words and the pattern of construction known, there is no limit to the production of other correct number names.

Because children enter school already knowing much about this despite the irregularities which pollute the verbal patterns, they are ready to deal with names of large numbers, some of which are large names, others of which are small – the name for instance 'million' is smaller than the name 'two-hundred-eighty-four'. Whether the children can correctly associate each name they produce with a particular model in pebbles is another matter. Fortunately this is not often needed at that stage. What is certain is that there is a great deal of number talk available with the basic vocabulary. Nearer and nearer approximations to perfection in knowing their meanings come – just as they do with the common language.

An interesting observation in this context is that in language arts curriculum guides it is rarely suggested that children be limited to words of two letters or one syllable in, say, grade 1, only to be allowed to speak and read longer words in the next grade, for fear that if that is the progression insisted on, then understanding will be clear and confusion avoided. Not at all. No limitations are suggested about the lengths of words of the common language, although that do es not imply the children do not have difficulties. No one suggests removing possible confusion by insisting that the length of words should control what teachers do with the students.

What then is this pattern for the production of number words?

Simply that the basic marks are, 0, 1, 2, 3, 4, 5, 6, 7, 8, and 9 (on Earth, but differently on 'other planets') and that number names beyond those are made up of combinations.

After 9 the next is 10, made up of a 1 and a 0. The 1 is then retained and the right-hand digit becomes successively that which followed the 0 in the basic vocabulary: 11, 12, 13, 14, 15, 16, 17, 18, 19.

Then 2 is substituted for the 1 in the lefthand position and on the right, digits are repeated in the same order as previously:

20, 21, 22 ... and so on, up as far as 29.

This pattern continues, as is well known, until all the possibilities of two digits are exhausted and three are needed. That's when 100, 101 ... and so on, follow.

None of that is new to any reader but is mentioned because the emphasis is here upon the simplicity of the pattern which can be seen with the eyes, even if no specific meanings are attached to every word; even if some of the meanings are in fact wrong, to be corrected at some later stage.

The spoken equivalents to the numeral words are, however, more irregular, patterns not being obvious for smaller numbers. If only after the word 'ten' the English language had used 'ten-one', or 'one-ten', or even 'one-teen', to be followed by 'two-teen', 'three-teen', fourteen, and 'five-teen'! But that can hardly be suggested as a widespread change nowadays, though teachers in school could suggest these as an unusual alternative which connects the meaning more easily with the words conventionally spoken.

Past 'twenty' the patterns settle down. They become consistent and reliable, and learners need less energy to become aware of large numbers or number names than they previously had to when no patterns were present.

Computational Efficiency

The secret of computational efficiency in arithmetic lies in the use of a small basic vocabulary of words which are used to make compound words – they are the standard names of numbers – and the art of moving to these from other kinds of compound words. The latter are the sums, differences, products and quotients and mixtures of these. Sums will be those which contain the plus sign, differences the minus sign, products the multiplication sign and quotients the division sign. Mixtures can include many combinations of the four signs.

Patterns will sometimes be visual – seen in the writings – or heard in the spoken language used. Or seen from the actions which the student makes upon some concrete materials. At other times the patterns will be neither visual nor aural.

The pattern seen in the construction of 10, 11, 12 . . . 19 is visual and is regular. That for the spoken English equivalents 'ten, eleven . . . nineteen' is aural and not quite so regular, though from 'twenty, twenty-one, . . . twenty-nine' the regularity is perfect. With sets of pebbles or with colored rods placed in a train, one can *see* the reversibility, giving the same total or combined length respectively.

However, when it comes to knowing whether or not $2 + 5$ in spoken words or in written numerals is an acceptable substitute for $4 + 3$, neither the written form nor the spoken provides any clue. One has to accept it from someone else who represents the authority of the local society. Even if concrete materials are referred to in this context, let it be noted that there is still need to be told something. One can neither know that 2 refers to ⊙ without being told the conventionally accepted name 'two' or '2', nor that the plus sign is associated with the act of putting together two sets.

Suppose on the other hand we know that $2 + 5 = 5 + 2$, either because we are told so or because we have developed this from sets. Without any more information there are clues present which can lead to a lot more.

We can see, or hear, that each is a reverse of the other. This strongly suggests that $4 + 3$ and $3 + 4$ are also equivalent and that reversability holds for the sum of *any* two standard names. We would be very surprised to find that our intuition in this were not justified.

To generalize the reversability principle for sums, therefore, little expenditure of learning energy is needed but it leads to a great deal, for even a young child will know at least one other equivalent for every sum. With no other demands necessarily made he can make up true sentences like this:
$$7 + 8 = 8 + 7 \qquad 200 + 500 = 500 + 200$$
$$3149 + 6728 = 6728 + 3149$$

This can be done without the student being able to read all the number names conventionally correct and he may have little understanding of the size of the numbers involved. The length of the compound nouns are not restrictions *in this situation* .

A few more examples may help to clarify:

a. The writing $2 + 5 = 7$ provides no clue as to its truth. Given this as true, however, and given also $20 + 50 = 70$ we have increasingly clueful situations. For it is now very likely that:

$$200 + 500 = 700$$
$$2000 + 5000 = 7000$$
$$20000 + 50000 = 70000 \text{ and so on.}$$

All the student has to do is notice the pattern of the 0's and he can develop the sequence as far as he wishes or a teacher requests. Again, the *size* of the numbers has nothing to do with his ability or inability to do this. Neither has the size of the number names.

A student may, of course, think from these clues that:

$$26 + 56 = 76$$
$$24 + 54 = 74$$

but they are not considered true by others. Clearly the complete set of clues is not yet available so a teacher, understanding why these 'mistakes' are made, will at a later time confront the learner with more clues and more embracing clues. Such 'mistakes', however, do indicate that the young student is basing his work on a search for patterns and that is *vital* to his understanding and progress.

An analogy in common language can be heard when a four year old frequently says 'taked, falled' using the pattern he has noticed for 'walked, talked'. When he notices that 'took, fallen' are used by his community he corrects himself, accepting more complex patterns.

b. 'Two plus five equals seven' is clueless but if one accepts it as true one can anticipate that

> 'two - ty plus five - ty equals seven - ty'
> 'two - hundred plus five - hundred equals seven - hundred'.

The first is not conventional but the second *is* acceptable by all members of the outside school community (at least those who speak English).

c. $19 = 7 + 12$ provides no clue as to why 19 is written in association with $7 + 12$. On the other hand, if one also accepts that $9 = 7 + 2$ there *are* clues to the likelihood of the truthfulness of $19 = 7 + 12$. A digit has been added to each of the 9 and 2.

The statement $19 = 10 + 9$ is perhaps not quite so clueless because we can see the marks 1 and 9 both on the left and right of the equals ($=$) sign. We might reasonably substitute 8's say, for the 9's and expect that $18 = 10 + 8$. The substitute 8 for 9 in $19 = 7 + 12$, gives no clue of what other substitution is necessay to give a true sentence, unless one knows that 19 is 1 more than 18 in which case it must be that $18 = 7 + 12 - 1$. That $29 = 20 + 9$, $36 = 30 + 6$, and so on – these follow because of clues noticed in $19 = 10 + 9$. Soon one accepts that, for example, $125 = 100 + 20 + 5$ and one can reliably forecast that:

$$238 = 200 + 30 + 8$$
$$6238 = 6000 + 200 + 30 + 8$$
$$56238 = 50000 + 6000 + 200 + 30 + 8$$

The spoken language also carries patterns and clues obtained by listening precisely to what is said.

All students can create such sentences for themselves even though they may not be able to read them in the conventional, 'correct' manner. They do not have to know fully the meaning of the large numbers. That's another story and can be reserved for future study.

From such examples we can see that a blend is available to students and teachers for developing and using the language of arithmetic; a blend of vocabulary invented from the basic words according to patterns. Some patterns are seen in the written symbols, others heard in the sounds of the speech. Still more become evident from experiences of actions, physical materials and the conversations about what is done. Except in the simplest of activities all such aspects will be present.

In general, it may be difficult either within ourselves or within students to spot precisely which aspect of the blend will be operating at any one instant. What one person sees as a clue another may not. What provides a clue to a learner on one occasion may not yield the same clue on another occasion. *But such detailed analysis is not necessary.* What is important is that every teacher and every student realizes that computation is neither just a matter of practising the conventions nor only one of discovering the basic principles. Even if it is appropriate to concentrate at times on one emphasis the end - product of efficient, quick, correct and enjoyable computation comes from greater awareness of the blend of conventions *and* discovery, with much experience of pattern usage.

A teacher's responsibility is to lead children to engage in work which ensures all this.

5

Some Beginnings of Computation

This is not the place to list in detail all the activities which are valuable to classrooms for the development of young children's arithmetic experience. A wide literature deals with this, as well as varieties of textbooks currently used in schools.

A brief description is important, however, to link what has been discussed so far with what will follow and to suggest how traditional exercises can be supplemented and extended according to the emphasis on language, pattern and infinity!

The longest number name I have ever done

Activities with 'Concrete Aids' and Discussion

Fingers

Many adults still believe the use of fingers for arithmetic should be discouraged. Otherwise children will come to rely on them and it is necessary for efficiency to do otherwise.

Our view is that fingers are available, both in school and at other times when the children can continue the study with them, and that fingers can provide personal models for entry into many mathematical and arithmetic topics. It is very unlikely the fingers will become a crutch especially when each student widens his (her) mathematical powers by going beyond what can be done on fingers. Such activities can also be the center of good conversation, between a teacher and the class, or between the children themselves in small groups.

Counting

The fingers of one hand or both, can be counted. That can go further than 'ten' if the fingers of several children are used. Perhaps classification is needed as to whether thumbs in this context are called 'fingers'.

It is not only the fingers, however, which can be counted. An interesting game, suitable for getting away from the traditional conditioned responses is for the teacher to hold up her hands, fingers extended, with the spoken challenge, "How many?"

"Ten" will probably be the first reply but with patience and stillness, which suggests that other replies are needed, many numbers begin to be named:

'Two'	('Why?' 'Because I was counting hands')
'Eight'	(spaces between fingers)
'Nine'	(spaces, including the one between the hands)
'Eleven'	(including the spaces outside the end fingers)
'Zero'	(rings – perhaps)
'Two	(little fingers, only)
'Eight threes plus two twos'	(or an equivalent of this, counting the 'sections' of each finger (3 each) and thumbs (2 each) . . . and so on.

The fact, of course, is that the hands are *not* numbers; they are objects. The explicit meaning of 'How many?' is, "What number *can* be associated with something you perceive when you look at this?"

Children can be encouraged to extend their imagination as far as they like, giving some sort of rationale for each number chosen.

Such an activity will not overthrow the cultural convention that the shape of one's hands in context will usually communicate 'ten'.

Order

The finger - counting can be done in different orders and provides another set of activities. Perhaps at first only four or three fingers are studied. The challenge is not only "How many fingers?" but "can you count them in a different order?" There is a maximum number of ways of doing this but it may be sufficient to elicit only that there are many different orders.

The various orders can be demonstrated though students soon begin to realize that it is difficult to keep track of them all. Finger - names are therefore assigned. "Thumb, first finger, middle finger." Or colored ink dots mark each finger leading to conversations like, "There's 'red, green, yellow', 'red, yellow, green' . . . and so on.

If the teacher knows the maximum number she will be guided as to when to press the students for other orders and when to ease up. In the case of 3 fingers, the orders discovered can be listed, just as with the change - overs in chapter 1. There are exactly 6 possibilities. For 4 fingers, each in turn can act as the first - mentioned to be followed by the names of the other three. But since we have found 6 orders for any three fingers, and each of the four can be in turn the first - mentioned, there must this time be 4 x 6 orders. For 5 fingers there will be 5 x 4 x 6, or to show the *pattern* more clearly, 5 x 4 x 3 x 2 x 1 . . . and so on!

Addition

A beginning to the 'addition of whole numbers' can be made by spreading the fingers of one hand, or both, in such a way that one can say,

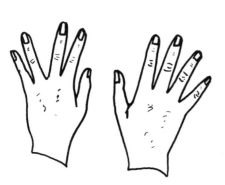

"Another name for five is 'two - plus - three'! What other names can you invent for 'five' using your fingers?" Answers include, 'Three plus two', 'one plus two plus two', 'two plus two plus one' . . . and so on.

Such activity can be accompanied by conversation, though written names can also be recorded by the students, numerals being used: 5, 3 + 2, 2 + 3, 1 + 2 + 2, 2 + 2 + 1. There can be a rich crop of results, since with both hands the numbers 1 through 10 are available and with other people's hands those greater than 10.

If 'zero' comes into some of the replies, well and good. There is then an endless number of possible names for each number, since we can include, for 4 say, 0 + 4, 4 + 0, 0 + 0 + 4, 0 + 4 + 0, 4 + 0 + 0, 4 + 0 + 0 + 0 . . . and so on. Adults may find such extensions excessively tedious because they do not need the practice. For children who come to this as a new experience it is usually challenging enough to maintain interest.

Subtraction

Fingers can be taken away from others. Beginning with five fingers, say, we can fold down two fingers. The children can call it, '5 take away 2 is 3'. Such an entry can lead to many other examples, $5 - 2 = 3, 5 - 1 = 4, 2 = 5 - 3, 3 = 5 - 2, 5 - 5 = 0$ . . . and so on.

Alternatively the phrase 'complement of' can be introduced into the conversation as the actions are made. "In 5, the complement of 2 is 3." "In respect to 5 the complement of 1 is 4." Or just "4 is the complement of 1" when, in context, everyone knows it is in relation to 5.

'Taking away' objects is one of the concrete models for the subtraction of numbers. Although the common usage of 'take away' is more appropriate for such finger activities, 'minus' and 'subtract' can also be used. They both can carry the meaning in this context although there will be other contexts in which 'subtract' will not mean 'take away' objects; 3 - 5, for example.

Addition and Subtraction

Any of the above mentioned work on addition and subtraction can be extended to have both signs included in one name or in one sentence. If $5 - 2$ is considered, then $4 + 1 - 2$ is an equivalent; $5 - 3 = 2, 4 + 1 - 2 = 3, 5 - 2 = 5 + 0 - 2 = 3 + 0$ . . . and so on, endlessly, beginning with any standard or non - standard number name suggested by the fingers.

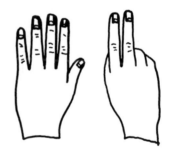

A particularly powerful example, arising spontaneously or from teacher's lead, is to bring one finger of one hand close up to the 5 fingers of the other. What elicited 5 now gives $5 + 1$. The lone finger is moved away. 5 is left. The actions suggest $5 + 1 - 1$ and that must be equivalent to 5.

If 2 fingers are moved close and then away, $5 + 2 - 2$ is seen to be also an equivalent for 5. Similarly for $5 + 3 - 3, 5 + 4 - 4$.

Likewise $10 = 10 + 1 - 1 = 10 + 2 - 2 = 10 + 3 - 3$ . . . and so on, by substituting 10 fingers for 5 fingers and repeating the actions, actually or virtually.

Extending the virtual possibilities students can imagine, say, 17 beads clasped in one closed hand. The other fist is moved close to it and then away. Maybe 6 beads are in this hand. Then $17 + 6 - 6 = 17$ and we can extend this with any number of examples. A few year's later, maybe, the formal writing will be met: $x = x + y - y$, but its understanding is important to early Arithmetic.

$5 - 3 + 3$ can also be investigated by folding down 3 fingers after stretching out five. The three are then unfolded, the original situation being restored. Thus, $5 - 3 + 3 = 5$ and the generalization follows as it did for $5 + 3 - 3$.

Multiplication

An introduction to multiplication can be begun in the context of the finger activity, although for young students it does not necessarily imply that a lengthy study of multiplication is about to commence. Simple examples, using the language, can be opportunely seen, valuable for what will follow.

Five fingers can be seen as 'Two pairs of fingers plus one finger'. $5 = 2 \times 2 + 1$ is written, or if preferred, as $5 = (2 \times 2) + 1$ to lessen the possible ambiguity that it might mean $5 = 2 \times (2 + 1)$.

Students, with a start like that, will probably produce $(2 \times 2) + 1 - 2 = 3$, $(2 \times 2) + 1 + 0 - 2 = 3 (5 \times 1) - 2 = 3$ ... and so on, again endlessly if zero is included.

Division

Division being the inverse of multiplication, it can also be gleaned from finger movements.

"How many groups, each of 3 fingers, are there in 10 fingers?" One can say, "I see one group of 3 fingers and 7 others" or, "I see two groups, each of 3 fingers, and 4 others" or, "I see three groups, each of 3 fingers, and 1 other".

The first recordings of division could be as on the left, though subsequently they will be written traditionally:

10, 3, 1, 7	$10 \div 3 = 1$ and 7 remainder
10, 3, 2, 4	$10 \div 3 = 2$ and 4 remainder
10, 3, 3, 1	$10 \div 3 = 3$ and 1 remainder

Correspondence

Another finger activity is to study the fingertips of one hand touching the fingertips of the other hand. Though the technical term, to be used later, is 'one - one correspondence' there is no point in stressing *that* . Instead, we can direct our attention to the awareness that we can look at the situation in two ways. Either as five groups (or 'sets') each of 2 fingers, or by a shift of the mind, as two groups each of 5 fingers. That is 5 twos or 2 fives; 5 times 2 or 2 times 5; $5 \times 2 = 2 \times 5$.

Also at hand are $4 \times 2, 2 \times 4; 3 \times 2, 2 \times 3; \ldots 1 \times 1$ and with the help of another person $6 \times 2, 2 \times 6; 7 \times 2, 2 \times 7; \ldots$ and so on. Products can be reversed, therefore, and without having to refer to the actual fingers an endless number of products can be named, written and reversed. The need to get the standard name for all such products is *not* present in this work though this will indeed be an emphasis at a later stage. Nevertheless a few will be known by first grade children: 'ten' for 'two times five', maybe others.

Summary of Fingers as Aids

The recommendation made is not for a wholesale use of fingers to develop all four basic operations and many families of equivalent number names. Only that because fingers are available to all children, they can act as models for study. When the numbers get past ten, the manipulations may become awkward if more than a couple of helpers have to pool their fingers.

It is the use of the few fingers to produce the beginning of the algebraic processes which is going to be so important in the ongoing learning of arithmetic and carried over to other branches of mathematics.

Counters

All the arithmetic and mathematics which was extracted from the actions on fingers can also, of course, be done with counters, pebbles or beads and probably should be so done to cater for individual tastes. What may be seen by one student studying fingers may only have been seen by another studying counters.

One advantage with counters is that it is easy to manipulate many more than ten. Hundreds of them can be made available for students to use.

Disadvantages include the fact that counters often move too easily and roll off the table! However, they are but models for the study of numbers and, as with fingers, students can *imagine* large numbers of them.

For example, "If in a box there are three hundred pebbles and you remove thirty - seven of them, how many are left in the box?

If I now replace those removed there are now three hundred again. Therefore, three hundred take away thirty - seven, put back thirty - seven, *must* be a name for three hundred." $300 - 37 + 37 = 300$

"Alternatively, if I put another forty - three pebbles in the box there are $300 + 43$ there. When I remove either the forty - three I added, or any other forty - three pebbles, 300 are again there." $300 + 43 - 43 = 300$

(Any appropriate numbers can be substituted, actually counted out, or, more powerfully, imagined.)

Fingers and Counters for Hundreds and Thousands

If children can count on their fingers, or with counters, 'one, two, three . . . and so on', they can also do so saying 'one hundred, two hundred, three hundred . . . and so on', regardless of whether at the time of doing so they may have any deeper understanding of hundreds. More understanding is certainly what we all want them to learn in time but these counting exercises can be practiced on the word level alone, perhaps extended also to the numeral code.

So '5 hundreds' can also be accepted as '1 hundred + 4 hundred', zero as '5 hundred - 5 hundred' . . . and so on, there being extensions using the word 'hundred' of every name and sentence previously generated.

If 'hundred' can be used, so can 'thousand', 'million', 'billion' and children love the sound of the large numbers. They can even use 'sevenths' and other fraction names, although they may not know what they mean.

In the case of the first few multiples of ten there are some irregularities. That's why 'hundreds' and 'thousands' were mentioned ahead of 'ninety', 'eighty', 'seventy', 'sixty'. But 'fifty' is irregular, not being 'five - ty'. So are 'thirty' and 'twenty'. 'Forty' is irregular in spelling, regular in sound. With these observations a finger can be called 'ten' and counting on the other fingers gives 'twenty, thirty, forty . . . and so on'.

Colored Rods

One of the first impressions most children have when they see a bag of colored rods spilled onto a table, especially if there are hundreds of rods, is that they are of different colors. Color - blind children, if they don't perceive that, will soon discover there are different lengths. That first impression is soon sharpened to an awareness of a certain number of colors or lengths. In short, they see *patterns*. When play with the rods begins the students find that all yellow rods are of equal length. So are the orange rods, blue rods, etc. These model other patterns, providing opportunity for an endless supply of 'equivalent lengths', 'non - equivalent lengths', sums, differences, products and quotients.

In the case of sums, shown by placing rods end - to - end touching, to form 'trains', a student will quickly know that an unlimited number of trains is possible for him, provided he has enough rods, provided he can get them on his desk, on the floor or on the Earth! When he becomes familiar with reading a few trains according to some code invented by himself or passed on from the culture by his teacher, he will realize he can also do so with *any* train of *any* length. It may take much energy, he may not *wish* to do so, but he will know that he can generalize for trains of any length. If the trains are so long that sufficient rods are not actually present, then the student can consider lengthening any train virtually, reading accordingly.

Should such activities be interwoven continually, appropriately and in socially acceptable classroom ways, with discussion and conversation then the three emphases examined in this book, language, patterns and infinity ('... and so on') are built in for the purpose of powerful mathematics learning.

While a sum is modelled by a train of rods, a difference is constructed by placing one train alongside another, longer, shorter or of equal length. A light - green rod, for instance, by the side of an orange rod shows a length difference. If order is not considered, the difference between the lengths is the length of a black rod. If order is held to be important in some context, there are two possibilities: o - g and g - o.

A product can be extracted when in a train there is repetition of color. That train which at first is read as 'red plus red plus red plus yellow', for example, written also perhaps as "r + r + r + y", can also be termed "three reds plus yellow", written as '3r + y' or '3 x r + y'.

A quotient arises also from a model of two trains, side by side. Such a model could of course suggest 'subtraction', but with a movement of the mind one can ask 'How many reds can be placed alongside an orange rod?' Correct replies include:

1 with a brown rod to complete the orange length,
2 with a dark green rod to complete the orange length,

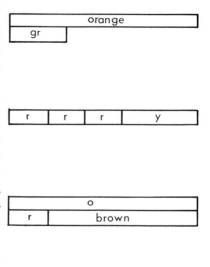

3 with a pink rod to complete
the orange length,

r	r	r	P

4 with a red rod to complete the
orange length,
5 with nothing to complete the
orange length.

r	r	r	r	r

Such informal oral reporting of what every student can do and see is most proper, though to allay doubts, perhaps we should add that later the more formal can come from the same actions:

$$10 \div 2 = 1 \text{ and } 8 \text{ remainder}$$
$$10 \div 2 = 2 \text{ and } 6 \text{ remainder}$$
$$10 \div 2 = 3 \text{ and } 4 \text{ remainder}$$
$$10 \div 2 = 4 \text{ and } 2 \text{ remainder}$$
$$10 \div 2 = 5 \text{ and } 0 \text{ remainder}$$

It is not trivial to include the last, with zero remainder, as later the pattern can be extended. Meanwhile, the last can be abbreviated to $10 \div 2 = 5$.

Number Names from Rods

There is a large repetoire of activities with rods which can provide excellent practice for early work in computation. One particular sequence of such activities is suggested by the development and exercises of chapter 3. Although there the context was within a base other than our standard ten, similar development can occur in any base. It is only parts of the basic vocabulary which differ from base to base. The mathematical processes remain constant.

Exercise

Repeat all the exercises of chapter 3 in another base. In particular use the common base. For that the orange rod will represent 10 when the white is the unit.
(On the 'blue planet', the blue rod is 10, the brown 8, black 7 . . . and so on.
On the 'brown planet', the brown is 10, the black 7, dark green 6 . . . and so on.)

Exercise

Which sentences will be true on several planets? Which false, which open?

Exercise

Will there be sentences true in all bases?
(e.g. $10 + 1 + 1 - 1 - 1 = 10$). Name others.

It is not suggested that children in primary grades necessarily become involved in the arithmetic of other bases, though there is evidence they find other bases no more difficult than the common one. If they do, however, the retention or memorization of standard names or of the tables, are not the objectives. The emphasis will be upon the patterns applicable which have analogies on our own planet. In the common base arithmetic we shall aim for the retention of names, non - standards and standards. We want children to know their addition and multiplication tables! They will be discussed in chapter 6.

A complicated arrow graph

Activities with Paper, Pencil and Discussion

Arrow Games

Arrow games are played on paper with pencils or colored felt pens. They provide practice in generating and using whole numbers and the operations upon them. One can proceed from a given standard name to various non - standards or vice versa.

Suppose 7 is written. An arrow drawn from it may represent 'suggests the equivalent name';
$$7 \to 6 + 1$$
$6 + 1$ may beget $1 + 6$; $1 + 6$ may beget $2 + 5$ and that $3 + 4$. There are so many alternatives for students to work at.

Some precision can be introduced at a teacher's discretion to the effect that one arrow stands for one mental change and one only.

Thus if someone writes $7 \to 6 + 1 \to 8 - 1$ it may be that in his mind the writer thought: "7 suggests $6 + 1$, $6 + 1$ suggests 7 and 7 suggests $8 - 1$". But that could more precisely be recorded as

$$8 - 1 \leftarrow 7 \rightleftarrows 6 + 1$$

If on the other hand the student says, "I know that if I added 2 to the 6 to get 8, then I had to subtract 2 from the 1 to get 8 – 1", then his representation by one arrow *is* precise.

Other details can be introduced. A transformation which changes a sum into its reverse can be shown by an arrow of a specific color, or shape if color is not used. A transformation which led to $8-1 \rightarrow 9-2 \rightarrow 10-3 \ldots$ and so on, would then have an arrow of different color or shape compared to the 'reverse' arrow, even though both kinds of arrow denote 'equivalent to'.

Several arrows, each standing for a specific kind of transformation can be drawn in the same diagram. Variations in going from one name to the next include,

equivalent to	1 more than	the square of
less than	1 less than	multiply by 3, then add 2
greater than	2 less than	. . . etc.

From such diagrams, students can extract any number of sentences, each linking two number names:

$$5 = 4 + 1 \qquad 3 \text{ is less than } 7 \qquad 7 > 3$$
$$1 + 6 = 7 \qquad 8 = 9 - 1 \qquad 17 = (5 \times 3) + 2$$

Symbols do not have to be conventional in all cases. The words 'less than' are available rather than the formal. 'The square of 3' can be written as ' $\square$ 3'.

Arrow games lend themselves to a great range of topics, for beginning number work and later for fractions, decimals, percentages and equations.

Two examples for equations follow. Here an arrow means 'is equivalent to', in the sense that if one equation is true then so is the second.

$$3 + 4 = 7$$
$$\uparrow$$
$$4 = 7 - 3 \leftarrow \quad 4 + 3 = 7 \rightarrow 7 = 3 + 4 \rightarrow 7 = 3 + 3 + 1$$
$$\downarrow \qquad\qquad\qquad\qquad \downarrow$$
$$4 + 3 = 7 + 0 \qquad 6 + 1 = 3 + 3 + 1$$
$$\downarrow$$
$$4 + 3 = 7 + 0 + 0$$
$$\ldots \text{ and so on}$$

$$y = g + r \rightarrow y = r + g$$
$$\uparrow$$
$$r + g = y$$
$$\uparrow$$
$$g + r = y \rightarrow g$$
$$= y - r \rightarrow y - r = g$$
$$\downarrow$$
$$r = y - g \rightarrow$$
$$y - g = r$$

(These letters can repre-
sent the lengths of a particular brand of colored

rods or, alternatively, any
numbers for which one of
the sentences is true.)

Different Symbol Systems

Some changes from the traditional, conventional symbols have already been suggested – in work on different bases and at other times. This is valuable experience for students in order that they do not become fixated on just one system of notation. The use of letters representing numbers contributes to this; sometimes a letter being the initial of a person's name thinking but not telling, of a specific number; sometimes a letter representing the length of a colored rod. A letter can also stand for one of the numbers arising from measurement of a rod. Although at first y for 'yellow rod' might persuade a young student to believe that y is a substitute for 5, the yellow length suggests many other numbers depending on the unit chosen. The letter y gradually assumes, therefore, the extended meaning of *any* number.

For some activities in equation transformation, students can use the initials of their names to denote any number they can think of. If P is Peter's number and he declares it to be 4 then $P = 4, P + 1 = 5, P + 2 = 6, 2 \times P = 8 \ldots$ and so on, thus generating members of a family of equations, all true if the first one is true.

If the sum of Mary's number and Ann's is written $M + A$ and if, further, it is assumed that $M + A = P$, Peter's number, then it follows that $A + M = P, P = M + A, P - M = A \ldots$ and so on, as previously for any true sentence of the same form using numerals.

When some students invent their own symbols they sometimes at first get carried away with the weirdness of the symbol shapes forgetting that they represent numbers. Such initial emphasis tends, however, to wear off under repeated challenges to the effect of, "Do your symbols represent numbers? If so are you sure that these equations are equivalent to each other?"

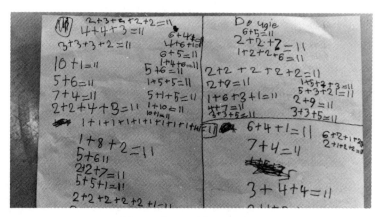

Names for 11--First grade

Mirror writings can also be tried, the *reflections* of which look normal. And a challenge now and again to do some arithmetic upside down can arouse keen interest! These too show variations of the symbols used.

Creating One's Own Exercises

Many of the above activities can lead to a requirement that students make up their own exercises. This can happen as a regular custom in lessons or be expected when work set by the teacher has been finished.

(a) Students devise other exercises, problems, examples, names, sentences of their own invention, or ones like those 'just finished'.

(b) Textbook exercises can be viewed as suggestions only, other exercises being provided by members of the class.

(c) Many individually - invented exercises in one lesson can be used for the whole class in a subsequent lesson.

(d) Students can be required sometimes to return to work done previously, correct themselves by referring to fingers, pebbles or rods, and by comparison and discussion with other students. New, perhaps better examples, can then be developed.

Another variation is to instigate some on - going written work bearing a title like 'spare time maths', 'hard maths' or 'fancy maths'. This provides opportunity for every student to write his own arithmetic/mathematics just as he might keep a diary with his own jottings. Such work does not have to be marked in detail, although teachers will want to celebrate with each pupil what she (he) has done, make suggestions and challenge.

Some examples of grade 3, 'super math'.

$$3247231145201 \times 14$$

$$\begin{array}{r} 12988924580804 \\ 32472311452010 \\ \hline 45461236032814 \end{array}$$

$$14 = \left(\left(\tfrac{1}{5} \times 200\right) - \left(\tfrac{1}{5} \times 200\right)\right) + 14$$

$$= \left(\tfrac{1}{7} \times 200 - \tfrac{1}{7} \times 200\right) + 14$$

$$= \left(\tfrac{3}{10} \times 999 - \tfrac{3}{10} \times 999\right) + 10 + 4$$

$$10 = \tfrac{1}{2} \times (9000000 - 2000000 + 2000000 - 9000000) + 2 \times 5$$

$$= 17^{\square} - 17^{\square} + 10$$

$$(600 + J) \div 17 = 6 \text{ and } (498 + J) \text{ remainder}$$

$$(A + J) \div B = C \text{ and } (A - B \times C + J) \text{ remainder}$$

At first, children may find this rather hard, perhaps not knowing quite what is wanted. It may be a departure from the traditional conditioning when teacher had to present all the exercises. Considerable patience may be needed before the students respond to the new responsibility. Some may think it a joke, others may find security at first in repeating over and over again the simplest sentences, 1 + 1 = 2, which they have known for a long time. Still others may copy from someone else before they come to realize they themselves can also take the responsibility for inventing arithmetic exercises.

Textbook Exercises

A plethora of paper and pencil exercises crowds the textbooks. Many of the so - called 'problems' can be useful in class, but usually they will need a richer treatment to fit in with the kinds of mathematics learning we seek in this book.

1) The exercises can be seen as reading matter alone, on which occasions fluency in reading is the aim, not the figuring out of problems or of getting answers.

2) Teachers and students can discuss how specific parts of a textbook could be improved.

3) Exercises from the book can be used to construct longer or different ones. e.g. if $2 + 3 = \square$ and $4 + 5 = \square$ are on a page, then $2 + 3 + 4 + 5 = \square$ is a new challenge, even if $5 + 9$ is the entry put in the box.

The main point here is that we are not recommending the abandonment of texts. They are materials to be used wisely by learners, under direction. Teachers must retain the professional obligation to use the text as they, themselves, best determine. This is not necessarily exactly as the publisher, or another teacher, has decreed.

Some worthwhile additions to paper and pencil work follow.

Using a Chart

A chart, on card or on the chalk board, displays a collection of number names and signs, selected according to the number - name vocabulary in current use by the students. A pointer is moved from chalk mark to chalk mark to form a new number name, or sentence if a sign $=$, $>$ or $<$ is tapped. The students read each when the pointer stops. Maybe they also write them down.

Alternatively, sentences can be composed in writing without the prior use of the pointer. All sentences do not have to be true. A task to follow later could be to determine whether each sentence is true, false, or 'not known'.

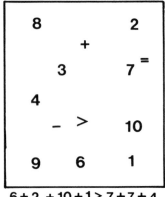

$6 + 2 + 10 + 1 > 7 + 7 + 4$

$39 = 8 + 10 + 10 + 10 + 3 - 2$

Different 'Box' Positions

There is common complaint that some students cannot complete open sentences. They cannot put a number name in a 'box': $3 + 4 = \square$. If this occurs, the children can be asked first to write *any* number name in the box, then commit themselves to write T, F or ? (True, False, Don't Know).

This applies to boxes in any position and to several boxes in the same sentence, opportunities again for creative replies:

$\square + 7 = 9$ $4 - \square = 2$ $3 + \square + 6 + 1 = 10$ $7 + \triangle = 3 + \bigcirc$

If several boxes are of equivalent shape they may imply that equivalent names should be entered. Or this can be disregarded. What matters is that one is aware of both possibilities and responds accordingly.

'Stretching' and 'Squashing' Games

These are variations of computational substitution activities. Below is shown one of each. The 'squashing' game does not *have* to end with a standard name, although the challenge is to squash as much as one can:

9	$1 + 2 + 4 + 2 + 3 + 1 + 2 + 10$
$1 + 8$	$3 + 4 + 2 + 3 + 1 + 2 + 10$
$1 + 6 + 2$	$3 + 6 + 3 + 1 + 2 + 10$
$1 + 2 + 4 + 2$	$3 + 6 + 4 + 2 + 10$
$1 + 2 + 5 - 1 + 2$	$3 + 6 + 6 + 10$
$1 + 2 + 6 - 1 - 1 + 2$	$9 + 6 + 10$
... and so on	... and so on

From One Number to Another

Two numbers are suggested. The task is to find many different ways for transforming one into the other. If written, collectively or individually, records could look like this:

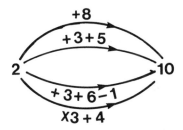 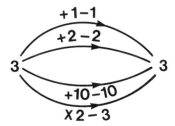

Restricted Vocabularies

Many of the foregoing activities can be varied by stipulating that 'in this game', only a restricted vocabulary is permissible.

e.g. Write number names, using 3's only, with + and - signs.
$(3, 3 + 3, 3 + 3 + 3, \ldots 3 - 3, 3 - 3 - 3, 3 + 3 - 3 + 3 - 3 \ldots$ and so on$)$.
Standard names are not necessarily demanded.

e.g. What are other names for $3 + 2$, using only 3's and 2's, and + and – signs?
$(2 + 3, 3 + 2 - 3 - 2 + 3 + 2 \ldots 3 + 2 - 22 - 22 + 33 + 22 \ldots)$

e.g. Note that arithmetic in each base *necessarily* implies a small restricted vocabulary (on the 'red planet' only 0's and 1's can be used).

e.g. If $1 + 2 = 3$ what else is true, if you can only use 1's, 2's and 3's?
($2 + 1 = 3, 3 - 1 = 2 \ldots 1 + 2 + 3 = 3 + 3 \ldots$)

e.g. Complete. Then make up more, using only 1, 0, 2 and 4.

$$\begin{array}{cccc} 2 & 12 & 102 & 1002 \\ +4 & +4 & +4 & +4 \end{array}$$

e.g. Write different sums using 2's, 8's, 1's and 4's only.
($2 + 8 + 1 + 4, 8 + 2 + 1 + 4, \ldots$)

Mostly products for 10

Problems

'Problems' in traditional arithmetic usually refer to those pieces of writings in non - arithmetic language which describe some applications of numbers to commerce, measurement, etc. A valuable lead - up to these is to have students regularly make up applications from the number names and sentences they study.

> e.g. if students know that $2 + 7 = 9$, they can be asked,
> "How could $2 + 7 = 9$ be used in a store?
> in measuring length?
> in playing with balls?
> in setting a party table
> . . . and so on."
> Such invented 'problems' can be talked about or written.

Summary

The above briefly described activities available for manipulative, oral and written practice for computational purposes, based on mathematical and learning principles, are appropriate for students of many ages. The examples given feature what can be done by primary students beginning the art of computation. Amendments can be made with longer names, longer sentences, greater expected output, fractions, decimals, percentages, 'problems'. Much of this will be the essence of following chapters.

6

The Addition Table
Its Patterns

When several true sentences involving addition and subtraction are known they can be collected together. It is not possible, in advance, to say which of these will have been produced or mastered by any specific student or class. Perhaps we can assure that for most children in the first grade $1 + 1 = 2$, $1 + 2 = 3$, $2 + 2 = 4$ and others will be known with confidence. Also, $3 - 1 = 2$, $4 - 1 = 3 \ldots$ and so on. Whatever specific individuals know a large number of these sentences can be listed from the whole class.

Instead of writing such sentences as in the previous paragraph, from left to right in sentences, they can be arranged in the form of a chart. There are two headings to the chart ranging from 0 through 10, though headings for rows and columns could be otherwise. One form of the chart numbers is to show the possible sums, compounds of the headings, thus:

+	0	1	2	3	4	5	6	7
0	0+0	0+1	0+2	0+3	0+4	0+5	0+6	0+7
1	1+0	1+1	1+2	1+3	1+4	1+5	1+6	1+7
2	2+0	2+1	2+2	2+3	2+4	2+5	2+6	2+7
3	3+0	3+1	3+2	3+3	3+4	3+5	3+6	3+7
4	4+0	4+1	4+2	4+3	4+4	4+5	4+6	4+7
5	5+0	5+1	5+2	5+3	5+4	5+5	5+6	5+7
6	6+0	6+1	6+2	6+3	6+4	6+5	6+6	6+7
7	7+0	7+1	7+2	7+3	7+4	7+5	7+6	7+7

Should this form of the chart be used first the learners will have practice in knowing which sum goes where and be more likely to associate a sum with its corresponding standard equivalent in the next form of the chart. This is where the standard names for the sums are the entries:

+	0	1	2	3	4	5	6	7	8	9	10
0	0	1	2	3	4	5	6	7	8	9	10
1	1	2	3	4	5	6	7	8	9	10	11
2	2	3	4	5	6	7	8	9	10	11	12
3	3	4	5	6	7	8	9	10	11	12	13
4	4	5	6	7	8	9	10	11	12	13	14
5	5	6	7	8	9	10	11	12	13	14	15
6	6	7	8	9	10	11	12	13	14	15	16
7	7	8	9	10	11	12	13	14	15	16	17
8	8	9	10	11	12	13	14	15	16	17	18
9	9	10	11	12	13	14	15	16	17	18	19
10	10	11	12	13	14	15	16	17	18	19	20

In a classroom the substitutions from sums to standard names may, of course, be done as a result of activities described in former chapters. When the chart is required the written substitutions can be effected by writing each entry, by erasing each sum in turn to be replaced by its standard, or small pieces of paper can be stuck over each sum as each student does his own chart construction.

Progress to further study comes by posing the question "What patterns do you see in the table?" There are many and it is not possible to say in advance which will be spotted or in what order. Many of them may have been seen already – used, in fact, to complete or check the final table.

Some of the common patterns noticed by children and adults now follow. They are indicated by description or by suggested exercises:

A. Alternative headings can be used. A table does not have to deal with sums from 1 to 10 only. Headings might be as follows:

  (i)　 0 - 12, common in some countries

  (ii)　 from 1 - 10, or 1 - 12

  (iii)　 even numbers only, 0, 2, 4, 6, 8, 10

  (iv)　 odd numbers only, 1, 3, 5, 7, 9

  (v)　 more numbers in one heading than in the other

Exercise A

Construct two tables for each of the heading alternatives listed above, one showing sums, the other standard names.

B. The entries in each row are repeated in a column. The entries in a column are repeated in a row.

Exercise B

What does this pattern imply about the number of 'addition facts' in the table which have to be remembered?

C. Using the compass points to identify parts of the table,

Exercise C

Look at the diagonals which stretch across the table from southwest to northeast. Is there a pattern? Will a similar pattern be present in every possible addition table? – those constructed in Ex. A, for example?

D. Every row (or column) is nearly identical to the one before and the one after.

E. **Exercise E**

Read the right - hand digit of each entry going down any column, or along any row. Compare the results in any row (or column) with those of any other row (or column).

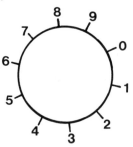

We can say that the digits are the same and in the same 'cyclic' order, as if they were written around a circle.

Beginning at any digit, go round either clockwise or counter - clockwise and one of the digit sequences is obtained.

F. The numbers in the main NW - SE diagonal are 0, 2, 4, 6, 8 . . . and so on. They are the standards corresponding to $0 + 0, 1 + 1, 2 + 2, 3 + 3$ . . . or $2 \times 0, 2 \times 1, 2 \times 2, 2 \times 3$ . . . and are called *even numbers*.

G. Study differences of entries in NW - SE diagonals.

Exercise G

Entries in the main NW - SE diagonal are:
0 2 4 6 8 10 ...

Differences between neighboring pairs of these entries are:

2 - 0   4 - 2   6 - 4   8 - 6   10 - 8   ...

or   2   2   2   2   2   ...

Find the differences likewise for any other parallel diagonals.

H. An addition table can also be used as a 'subtraction table'. We say, for example, 'seven minus three', or 'seven subtract three' and write 7 – 3.
7 is located in column 3 (or row 3) and for the heading of its row (or column) there is 4.

So 7 – 3 = 4

Exercise H

Find or check the standard names for many differences, using the addition table.

Many other patterns can be discovered. The others reported here may not be met if all a student wants is to memorize the most common patterns.

I. Suppose we begin with an entry, *any* entry, and move from it to another entry according to the moves, 'one step East followed by one step South', (1E, 1S) for short.

e.g., begin with any entry 4. A move (1E, 1S) takes us to 6. We can say that 4 is transformed into 6.
5 becomes 7, 3 is transformed into 5 . . . and so on.

Ex I$_1$ Trace some of the transformations with the move (2E, 1S).
Ex I$_2$ Trace some of the transformations with the move (3E, 2S).
Ex I$_3$ Make up another move and use it similarly on the table.
Ex I$_4$ Given any entry to begin with, can the transformations be predicted without consulting the table?
Ex I$_5$ What happens with moves like (3W, 2N), (4W, 9S) . . . etc.?

J. Imagine any rectangle, its sides parallel to the rows and columns, superimposed on the table to enclose some entries.
From such a rectangle (which includes a *square* rectangle) extract the two sums formed by the numbers in opposite corners.

6	7	8
7	8	9

6 + 9 and 7 + 8
are equivalent . . .

10	11	12	13	14
11	12	13	14	15
12	13	14	15	16
13	14	15	16	17
14	15	16	17	18

. . . so are 10 + 18 and
14 + 14

0	1
1	2
2	3
3	4
4	5
6	7
7	8

$$0 + 8 = 1 + 7$$

Exercise J

Check that this pattern holds for every possible rectangle of the standard table.
Compare each rectangle with the corresponding rectangle in the sum form of the table. What are the sums of the pairs of opposite corner entries *now*?

K. An interesting 'tapping game' can be played on the table which may not reveal anything new but may strengthen what has already been seen.
For example, using a pointer, a teacher can select in row 3, the entries 9 and 7. She taps the 9, then taps the 7, then moves to tap in succession the 8 and the 6 in the row immediately above.

She says, the words coinciding with the taps: "another name for 9 minus 7 is 8 minus 6. Who can tap out another name for 9 minus 7?"

A student will probably move up one more row with the pointer, tapping 7 and 5, then perhaps 6 and 4 up another row . . . and so on. Alternatively, the pointer can be moved downwards to a corresponding pair.
A family of pairs is established, each corresponding to a pair of taps:

(9, 7)   or written 9 – 7
(8, 6)   8 – 6
(7, 5)   7 – 5
(6, 4)   6 – 4 . . . and so on.

This *family* of pairs may also be indicated in some way like this:

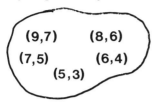

It has an *endless* number of members.
(2, 0) is *one* of its members.
The standard name of the family is 'positive 2', + 22.

Alternative writings are:

$$9-7 = 8-6 = 7-5 = 6-4 = 5-3 = 4-2 = 3-1 = 2-0$$

L. Another tapping game involves *three pairs* of taps, each time in the same row!

For example:
In row 1:'tap 6 tap 4   tap 4 tap 3
                     pause
                tap 6 tap 3
Three numbers, *any three,* in a row are selected and tapped as indicated.
Each is tapped *twice* , six taps in all.
The middle number is tapped twice successively.
The other two are also tapped twice, one at the beginning, one just *before* the pause, and then these are repeated *after* the pause.

Young students are able to learn this, without the teacher telling them. All they need to do is to look and listen (to the tap sounds), to learn what is called for in 'the game'.
When everyone can play the game in any row (or any column), records can be made.

The example given above is written,
$$(6-4) + (4-3) = (6-3)$$
An endless supply of examples is available.
$$(row\ 1)\ (11-8) + (8-3) = 11-3$$
$$(row\ 4)\ (9-6) + (6-0) = 9-0$$
$$(9-6) + (6-1) = 9-1$$

In fact, of course, there is no need to seek new examples in another row, for all examples are in *any one row* (or column).
The pattern here is the transformation of the long number name (on the lefthand side) to the shorter, by omitting the middle number, that which was tapped twice in succession.

One could, or course, resort to standard names.
$$(9-6) + (6-1) \qquad\qquad 9-1$$
$$= 3 + 5 \qquad\qquad = 8$$
$$= 8$$
so that $(9-6) + (6-1) = 9-1$, but the 'cancellation' involved provides a direct insight into the structure beneath. Later this may be more formally expressed in some way like this:

$$(a - b) + (b - c) = a - c$$

M. As noted already, in each column (or row) the numbers increase by 1 as we trace them downwards (or eastward) on the table. Reversing this, we can see the numbers *decrease* by 1 each time, if we begin low down and move up – 'northward'.

e.g., in column 2 ... 7, 6, 5, 4, 3, 2

What is the next number in this decreasing sequence? Surely it must be 2 - 1 or 1 and can be entered *above* the heading 0 as follows.

-4			-2		
-3			-1		
-2			0		
-1			1		
+	0	1	2	3	4
0			2		
1			3		
2			4		
3			5		
4			6		
5			7		

What comes after the 0, if we accept there is such a number? Surely 0 – 1, read "zero minus one." Then 0 – 2, 0 – 3, 0 – 4, ... and so on. Every column (and row) can be extended in this way. Perhaps '1 below' can be used by students as a temporary standard name for 0 - 1, analogous to its use in temperature scales. 1B or 1b could also be written. Alternatively, the official standard 'negative 1' can be introduced with no special feeling at this stage that this is to be preferred. − 1 is the standard writing for 'negative 1'.

When all columns and rows are extended (the headings too) we have a table that can be continued as wished in all four directions.

-8	-7	-6	-5	-4	-4	-3	-2	-1	0	1	2	3
-7	-6	-5	-4	-3	-3	-2	-1	0	1	2	3	4
-6	-5	-4	-3	-2	-2	-1	0	1	2	3	4	5
-5	-4	-3	-2	-1	-1	0	1	2	3	4	5	6
-4	-3	-2	-1	+	0	1	2	3	4	5	6	7
-4	-3	-2	-1	0	0	1	2	3	4	5	6	7
-3	-2	-1	0	1	1	2	3	4	5	6	7	8
-2	-1	0	1	2	2	3	4	5	6	7	8	9
-1	0	1	2	3	3	4	5	6	7	8	9	10
0	1	2	3	4	4	5	6	7	8	9	10	11
1	2	3	4	5	5	6	7	8	9	10	11	12
2	3	4	5	6	6	7	8	9	10	11	12	13
3	4	5	6	7	7	8	9	10	11	12	13	14

N. New sums can now be extracted from the extended table. $^-1 + {}^-2$, apparently, from entries in the NW part of the table, has a standard name of $^-3$. Therefore, $^-1 + {}^-2 = {}^-3$
and similarly from the use of other headings:

$$4 + {}^-1 = 3$$
$$^-3 + 2 = {}^-1 \ldots \text{and so on.}$$

If desired one could transform each of these sentences involving sums into sentences about differences:

$$^-1 + {}^-2 = {}^-3 \to {}^-3 - {}^-2 = {}^-1$$
$$4 + {}^-1 = 3 \to 3 - {}^-1 = 4$$
$$^-3 + 2 = {}^-1 \to {}^-1 - 2 = {}^-3 \ldots \text{and so on.}$$

Further, patterns can be spotted in the symbols used so that standard names may be processed without having to consult the table directly.

e.g. $^+4 + {}^-1 = {}^+3$ and $^+3 - {}^-1 = {}^+4$ (The $^+$ sign can be used to distinguish clearly between 4 and $^-4$.
also $^-1 + {}^+4 = {}^+3 \to {}^+3 - {}^+4 = {}^-1$

Any sum is reversible in the extended table, a difference is not. But a difference can be transformed into a sum.
(Note: The formal rules may not be easily grasped at this stage though they are difficult only in the large number of cases to examine, viz: $a + b$, $a - b$ where a and b can be 'positive' or 'negative' and $a > b$ or $a < b$.).

O. Now that the addition table has been extended to show positive, zero and negative numbers, we can study which of the patterns discovered previously, also apply to the extension.
Those described in B and C certainly do. Pattern D holds also though with our awareness of the possible extensions for every column and row in two directions we would qualify our meaning of 'the beginning' and 'the end'.
Patterns, E, F, G, H, I, J will also be seen to hold good.
For the game of I, in which a coded double move such as (2S, 3E) produced a sequence of entries, we can now move N and W without coming to the edge of the table. Transformation patterns can include (3W, 2N), (4W, 9N) . . . and so on.
For J, rectangles can be drawn, or imagined, completely surrounding negative entries or which contain both negative and positive numbers, with perhaps some zeroes. In the latter case some of the headings may be included but they do not interfere with the pairs of sums of numbers in opposite corners of the rectangles.
Corresponding to the pattern K more members of the family 'positive 2' can now be extracted from the extended table, such as $(1, {}^-1), (0, {}^-2), ({}^-1, {}^-3) \ldots$
We can also write, $9 - 7 = 8 - 6 = \ldots = 3 - 1 = 2 - 0 = 1 - {}^-1 = 0 - {}^-2 = {}^-1 - {}^-3 = \ldots$
The tapping game in L also has more scope because there are positive and negative entries to be tapped.

In row 1: $(6 - {}^-2) + ({}^-2 - {}^-5) = 6 - {}^-5$
In row 4: $(9 - 6) + (6 - {}^-1) = 9 - {}^-1$

All sentences of this form can again be summarized as,

$$(a - b) + (b - c) = a - c$$

where a, b, c this time represent positive or negative numbers b or zero.

P. All the patterns so far have been gleaned from the numeral system using the common base.

Previously, however, in chapter 3 an excursion was made to the 'pink planet', so if we wish, the addition table for that base, base IV, can be used for all studies we have made in this chapter.

The addition table will look thus:

+	0	1	2	3	10
0	0	1	2	3	10
1	1	2	3	10	11
2	2	3	10	11	12
3	3	10	11	12	13
10	10	11	12	13	20

All the patterns B, C, D will still hold good; E also, though the entries in the cycle will be . . . 0, 1, 2, 3, 0, 1, 2 . . .

Memorizing, say, the even numbers in this base will not be important to members of this Earth planet, though who will claim this may not be vital to the arithmetic on the 'pink planet'? It may, however, be most important to members of all planets to be aware that there *are* even numbers which exist independently of the symbols which happen to be used for them on any one particular planet! And this applies to many other properties of numbers.

For reference, the addition tables are shown in Bases II, III, V, VI, VIII and IX:

Base II — 'red planet'

+	0	1
0	0	1
1	1	10

Base III' — 'light green planet'

+	0	1	2
0	0	1	2
1	1	2	10
2	2	10	11

Base V — 'yellow'

+	0	1	2	3	4
0	0	1	2	3	4
1	1	2	3	4	10
2	2	3	4	10	11
3	3	4	10	11	12
4	4	10	11	12	13

Base VI — 'dark green'

+	0	1	2	3	4	5
0	0	1	2	3	4	5
1	1	2	3	4	5	10
2	2	3	4	5	10	11
3	3	4	5	10	11	12
4	4	5	10	11	12	13
5	5	10	11	12	13	14

Base VII 'black'   Base VIII 'brown'

+	0	1	2	3	4	5	6
0	0	1	2	3	4	5	6
1	1	2	3	4	5	6	10
2	2	3	4	5	6	10	11
3	3	4	5	6	10	11	12
4	4	5	6	10	11	12	13
5	5	6	10	11	12	13	14
6	6	10	11	12	13	14	15

+	0	1	2	3	4	5	6	7
0	0	1	2	3	4	5	6	7
1	1	2	3	4	5	6	7	10
2	2	3	4	5	6	7	10	11
3	3	4	5	6	7	10	11	12
4	4	5	6	7	10	11	12	13
5	5	6	7	10	11	12	13	14
6	6	7	10	11	12	13	14	15
7	7	10	11	12	13	14	15	16

Base IX 'blue'

+	0	1	2	3	4	5	6	7	8
0	0	1	2	3	4	5	6	7	8
1	1	2	3	4	5	6	7	8	10
2	2	3	4	5	6	7	8	10	11
3	3	4	5	6	7	8	10	11	12
4	4	5	6	7	8	10	11	12	13
5	5	6	7	8	10	11	12	13	14
6	6	7	8	10	11	12	13	14	15
7	7	8	10	11	12	13	14	15	16
8	8	10	11	12	13	14	15	16	17

All the patterns can again be re - searched, just as they have been in Bases X and IV:

> subtraction; the right hand digits in entries in rows and columns; extension to negative number names; symmetries; 'differences' from NW ↔ SE diagonals; families of differences, etc.

Q. There is still another way of transforming the number names in the addition table.
Suppose for any standard name, its digits are added, the same operation being repeated on *that* result until finally a single digit is reached.
For example, in Base X, the common base:

> 13 will be replaced by 1 + 3 or 4.
> 89 will be replaced by 8 + 9 or 17 and that by 1 + 7 = 8.

The results for all the entries in the 9 by 9 table are shown;

+	1 2 3 4 5 6 7 8 9
1	2 3 4 5 6 7 8 9 1
2	3 4 5 6 7 8 9 1 2
3	4 5 6 7 8 9 1 2 3
4	5 6 7 8 9 1 2 3 4
5	6 7 8 9 1 2 3 4 5
6	7 8 9 1 2 3 4 5 6
7	8 9 1 2 3 4 5 6 7
8	9 1 2 3 4 5 6 7 8
9	1 2 3 4 5 6 7 8 9

This is sometimes known as the 'Vedic' square, from the ancient Hindu sacred books called the Veda. It was thought to have magical properties.
Patterns can easily be spotted: symmetries; repetitions of rows and columns; more repetitions of some entries than in the regular table.

Each entry appears nine times, once in every column, once in every row. Except for the 1's each digit is used for the whole of two NE ↔ SW diagonals. If a digit (4, for example) appears 3 times in one of these diagonals it appears in the other diagonal, the 'complement of 3 in 9' times. Namely, 6 times.

6 is the complement of 3 in 9
7 is the complement of 2 in 9 . . . and so on.

Approach

Though we have listed the patterns one after the other, without apparent interruption, the implication must not be taken that it will occur with a group of learners, in exactly the same way.

Lessons on 'patterns in the addition table' can punctuate much other work on Arithmetic or Mathematics over many months and it is not possible to forecast precisely what patterns will arise nor the order in which they do. Also relevant in this will be the particular lead the teacher takes on each occasion and what patterns she decides not to stress until later.

Referring back to chapter 1 ('the approach') we can see how the various items apply again.

1. The entry to the activity is the construction of the table showing sum names, followed by the second table substituting standards for sums and the challenge to find patterns.

2. The limited situation is, of course, the table as presented and seen by

every student, whether or not each is capable of reproducing the whole thing for himself.

3. There is opportunity for any student to spot patterns that no one else in the class has, or to record on paper as many entries as wished from the 'infinite extension' of the table. Every student can see *some* pattern, no matter how simple, yet there is enough potential in the situation for anyone to be challenged more and more.

4/5 The pattern seeking is best done in a climate of discussion, each student accepting a commitment to say to the others what he sees, no matter how tentative. What some say will help others see.

6/7 Important, too, is the use of notepad and pencil. The children should be encouraged to write and try things as they go, reproducing a few entries for themselves or marking some on a dittoed copy of the table, finding differences when appropriate or recording a family of pairs.

9. Some people will think they see a pattern, only perhaps to change their minds after further investigation. Tentative beginnings, patiently accepted by others, frequently lead to joyful 'aha's!

10/11 Networks of inter - relationships will grow during the weeks of working on the table:

a) a continuation of the strategy of 'patterning' essential to both mathematics and to feelings of self - esteem.

b) awareness that in the innocuous looking addition table there lies a feast of information.

c) a great deal of practice in using standard names corresponding to sums and differences. This surely ensures that many will be remembered.

The mathematics

Many of the awarenesses coming from these patterns are important to students studying mathematics, although the emphasis here has been intuitive and not of a rigorous type of definitions and procedures. Referring back to the patterns as listed previously:

B) The fact that 'sums can be reversed' is technically said to be because of the 'commutative law'.

C) This pattern follows from the choice of headings. If one name in a heading is a, the next is a + 1; if another is b, the next is b + 1.

+	a	a + 1
b	b + a	b + (a + 1)
b + 1	(b + 1) + a	(b + 1) + (a + 1)

The entries in the SW ↔ NE diagonal are:

$(b + 1) + a$ and $b + (a + 1)$

But $(b + 1) + a = b + 1 + a$
$= b + (1 + a)$ associative law
$= b + (a + 1)$ commutative law

D) This follows from C.

E) The numeration system was purposely developed such that a righthand digit is repeated after ten entries.

F/G The differences between every adjacent pair of entries in any
NW ↔ SE diagonal is always 2.

+	a	a + 1
b	b + a	b + (a + 1)
b + 1	(b + 1) + a	(b + 1) + (a + 1)

Example: $(b + 1) + (a + 1) - (b + a)$
$= b + a + 2 - (b + a)$
$= 2$. (for all a, b in the headings)

H) This observation, which is true for any two numbers whose sum is an
entry, can lead to this generalization:
$$a + b = c \rightarrow c - b = a$$
"a plus b equals c' implies that 'c minus b equals a".
or "if c is the standard name for a + b, then a is the standard for c - b".
a, b, c are any numbers such that a + b = c.

I) Generally, we can see if we begin at any entry, a + b, and take, say
ℓ steps to the East (ℓ being the name of some number) and n steps to
the South, then we go from a + b to (a + ℓ) + b, and finally reach
(a + ℓ) + (b + m) or (a + b) + (ℓ + m).
We can write (ℓ E, mS) (a + b) → (a + b) + (ℓ + m) to represent
all this in a shorthand manner.

J) Again, let a and b represent two entries in the same row of the table,
any two.

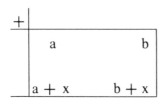

Consider a rectangle such that a and b
are located at two of its corners.
Suppose in the same column as a we
choose another entry a + x. The rectangle can be completed. Its fourth corner
must be at the entry b + x, since it must
be below b the same distance as a + x is
below a.

Sums of opposite corner entries are now a + (b + x) and (a + x) + b.
But a + (b + x) = a + (x + b)
$$= a + x + b$$
$$= (a + x) + b$$
The sums of opposite corner entries of every rectangle are, therefore,
equivalent, for our proof does not depend upon the choice of a, b or x.
They can be any numbers of the table.

K/L/M) Any later, more formal, definition of positive and negative
'integers' can now take into consideration this extension of the addition
table. If one wants this definition to harmonize with the table then if a is
any entry in the regular table, we shall require ⁻a to be the standard
name for 0 - a.
This, of course, is exactly what is done.
Moreover, the integers can be a little more rigorously defined thus:

The set of *all* ordered pairs of whole numbers has an endless membership list:

(0, 0) (0, 1) (0, 2) . . . (1, 0), (1, 1), (2, 1)

(3, 1), ... (15, 87), . . . (87, 15) . . . (101, 2) . . .

Certain subsets of this set will be called integers, either positive or negative integers.

Corresponding to any one ordered pair of these numbers, call it (a, b), we choose the *subset* of all ordered pairs (x, y) for which a + y = b + x. This set we call the 'integer (a, b)', perhaps shortening this to I(a, b). Note: (a, b) is a pair, (b, a) is *another* pair. The order is to be stressed, so we call it an 'ordered pair.' I(a, b) is a *set* of pairs.

Example:

The integer (7,11) or I(7,11) has the ordered pair (8,12) as a member, because:

$$7 + 12 = 11 + 8$$

Other members of I(7,11) are:

(9,13) for 7 + 13 = 11 + 9
(10,14) for 7 + 14 = 11 + 10
(11,15) for 7 + 15 = 11 + 11
(12,16) for 7 + 16 = 11 + 12
(0,4) for 7 + 4 = 11 + 0
(98,102) for 7 + 102 = 11 + 98

There are countless other ordered pairs that are members of I (7,11), and countless other ordered pairs that are NOT members!

Conventionally, this integer is written ⁻4; its standard name is "negative four".

This later development is now sensed in the family of pairs chosen in the tapping game:

(9, 7)	or written	9 – 7
(8, 6)		8 – 6
(7, 5)		7 – 5
(6, 4)		6 – 4
(1, ⁻1)		1 – ⁻1
(0, ⁻2)		0 – ⁻2
(⁻1, ⁻3)		⁻1 – ⁻3 etc.

This integer is + 2, "positive two."

N) Since the extended parts of the table are now seen to be developments from the original part and we have chosen our meanings to harmonize, all the underlying patterns we discovered in the regular addition table will now hold for the new entries.

O) This later may lead to a more formal, step by step proof, that for any three numbers a, b, c of the extended table:

$$(a-b) + (b-c) = (a-b) + b-c$$
$$= \{(a-b) + b\}-c$$
$$= \{a-b + b\}-c$$
$$= a-c$$

P) All patterns of numbers, regardless of the way they are represented, will exist in the *meanings* inherent in any of those representatives.

Patterns B) C) D) are like this.

Pattern E, however, is a pattern about one particular form of representation. There will be still a cyclic order present but the symbols in base IV only extend from 0 to 3. Similarly, in the tables of other bases, the cyclic order pattern will be seen, though there will be slight name adjustments according to what is permissible in the particular base.

7

Addition and Subtraction
Without Borrowing

Much of the emphasis so far has been on the awareness of patterns, either in the substitution of number names, the substitution of sentences or of various kinds in tables. These learnings have been considered basic and essential, being more important than the traditional task of emphasizing standard names or 'answers'.

Nevertheless, standard names have, of course, punctuated much of what we have done. Even when we have used non - standards they have been composed of other standards. For standards can be seen as simple nouns, while non - standards are compound names, compounded of some of the simple nouns. Standard names are, therefore, inescapable!

We have also been content with the writing of number - names which are made from left to right, because that is the common convention for our words and sentences when written.

There is need, however, for children also to study efficient substitution techniques to produce standard names for numbers which are large and associated with written standards using two, three or more digits. Traditionally, the methods used for these are called 'algorithms': arrangements of numerals set out horizontally and vertically which allow for procedures to follow more easily by virtue of the arrangements.

Addition

In addition, for example:

$$
\begin{array}{r} 278 \\ 342 \\ + 693 \end{array} \quad \text{and} \quad 278 + 342 + 693
$$

are possible. In the second layout the eye has to jump from 3 to 2 to 8 (or in reverse) to add the units figure. In the first arrangement that still has to occur but there are no intervening marks to interfere.

Traditionally, the first arrangement was the dominant and was taught rote - fashion. One began invariably on the right and "learnt the patter:"

> "3 and 2 make 5, and 8 makes 13.
> Put down 3 and carry 1." (Though it is *ten* that is meant).
> A figure 1 was usually written above the top of the next column to the left.
> "9 and 4 make 13, and 7 – that's 20 and the 1 carried, makes 21.
> Put down 1, carry 2" (though this represents two *hundred*). 2 is written at the top of the next column to the left.
> "6 and 3 make 9, and 2 is 11, and the 2 carried gives 13."
> Answer: 1313

This is quite satisfactory, of course, and provided one applies the algorithmic techniques correctly the right standard name answer is ensured.

Our view, however, is that this is only *one* possible procedure which can be, and should be, gleaned from the pattern development. The converse unfortunately does not so easily occur, namely that the powers of mathematizing do not commonly evolve when the thrust is only that of 'getting the right answer' in one way.

Let us study, therefore, what are the alternatives in the context of larger numerals.

First, note that even in the traditional algorithms, numbers are added only in the sense of what is done with a set of substitutions. One still transforms a 'word' "3 + 2" into another "5" and so on, until the standard form is reached.

In regard to the alternatives of horizontal or vertical arrangements - or even the numerals scattered at random over the paper - the place - value patterns have still to be honored. So we can study the advantages and disadvantages of each display and get used to the forms used by other people and in books.

Given such prerequisites we can proceed:

a)   We do not *have* to begin on the right. We can go from the left.

278				
342				
693	6	(hundred) + 3 (hundred)	→	9 (hundred)
1100	9	(hundred) + 2 (hundred)	→	(hundred)
200	9	('ty') + 4 ('ty')	→	13 ('ty')
	13	('ty') + 7 ('ty')	→	20 ('ty')
13				
1313		'adding up' or 'down', the standard name is 1313		

b)   We can begin in the middle.
c)   In any column we may move up or down.

These processes incorporate the property of addition technically called 'the principle of commutativity', and perhaps the 'principle of associativity'. Commutativity refers to the legitimacy of reversing a compound number using only + signs; associativity to the acceptance that, for example, $(3 + 2) + 8 = 3 + (2 + 8)$. Advanced students might be challenged to identify in a set of procedures as the example above, exactly when each principle is used. But it is not necessary for computational purposes. (Some aspects of 'New Mathematics' tended to overdo this). All one's energies are needed for the substitutions.

d)   If one can look up or down the columns, the line of sight being

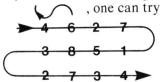

, one can try

This path is interesting as one can accumulate as one proceeds. There is no need to write 'carrying' figures nor subtotals.

"Four thousand six hundred twenty seven, 4628, 4678, 5478, 8478, 10478, 11,178, 11,208, 11212."

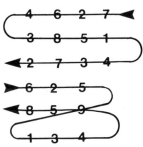

The arrow provides an alternative path for one's eyes but is more tricky because one has to avoid the 5 in the middle on passing it the second time. Nevertheless, this is worth a try, to challenge students as to possibilities, to test their skills at keeping track of what they are doing, and to help them appreciate how difficult some alternatives are!

e)   "Long tots" are vertical algorithms which involve a dozen or more numbers. Though computers may be used in commercial situations for totalling such examples, there is pleasure to be derived from tackling some in school. They can be invented on the spot by the students although recourse to the lists of numbers in any telephone book will provide more than enough.

42
73
169
62
204   550

38
453
67
2
942   1502

38
17
49   104

2156

The only advice needed is not to aim at standardizing the whole of each column in one substitution. Sub - totals can be noted and subsequently dealt with to get the final standard. They are not necessary, of course, but the experience of being interrupted halfway up a long column usually convinces one that writing the subtotals is wise!

There is one other advantage of becoming a master of alternative strategies. One does not have to rely on external authority to say that an answer is correct. One *knows*, because one can check by alternative strategies. Repeating the same strategy may well cause a repetition of the same error. A different attack will probably avoid any initial slip, give a different answer and a third check will be needed.

With such mastery any number of numbers can be added and the youngest children in school can at least attempt any example provided the degree of speed and accuracy is stressed only gradually over years. At first the emphasis is on basic patterns, on devising appropriate substitutions ("combination" is a modern word in this context), venturing on to longer numerals even if the answers are not always those sophisticated standard names of fellow students or teacher.

Gradually, the accuracy and speed are increased, less written steps are needed, "carrying" figures omitted. This emphasis, however, may not be wise to stress before the late elementary grades or even later when the adolescents begin to appreciate the slicker and quicker computations for the working world or for further study.

Subtraction

While the traditional strategies for addition seem to be retained by most children and adults, enabling them to produce fairly satisfactory accuracy, many children – according to their teachers and formal assessment studies – find subtraction more difficult. Except in the cases where each digit in the second number is less than or equal to its corresponding digit in the first, the strategies remembered do not always give assurance of a correct result.

The main method still used depends upon 'borrowing'. This strategy arose a century ago when thousands of well - intentioned adults tried for the first time to help children get right answers in subtraction. Many succeeded, but other children hardly mastered the processes involved and did not understand what was done or why.

In effect, the Victorian teachers said, "If you can't subtract, borrow." If, for example, the 'problem' was 531 - 174 the children were told to say "4 from 1 won't go, borrow ten". What the students did with their pencils was to cross out digits and insert others:

$$
\begin{array}{ccccccc}
531 & 531 & 531 & 531 & 531 & 531 & 531 \\
-174 & -174 & -174 & -174 & -174 & -174 & -174 \\
 & & & & 7 & 7 & 57 & 357
\end{array}
$$

There were some slight variations.

Now this is perfectly legitimate, though these days we would add 'provided that at the back of all this writing there is security' – meaning that if one loses one's way in the middle of the calculation one can clear the air and begin again, perhaps with a different attack.

"Are there alternatives to this rote learned routine?" can be asked. There are, undoubtedly!

To begin with we can understand the numbers cannot be borrowed. At least they cannot be borrowed in the same way as, for example, pencils can be borrowed. When Tommy borrows my pencil *he* has it and I *don't*. when he 'pays me back' he does not have it. I *do!*

But when Tommy says, "Loan me a number" or "Give me a number", I *still* have the number! In fact, no one can take away the numbers I have. I cannot lend them. I cannot have them borrowed.

Not that one should necessarily stop using the word 'borrow' in this context. It is not what word is used that is important but what other people might be meaning by our word.

Suppose, therefore, instead of trying to memorize gimmicks of symbol manipulation and explain something with words which would never have been used if the nature of numbers and subtraction had been appreciated; suppose instead of all that we dealt with the basics and then tried to see if children could be easily introduced to them? What then?

First, let us remind ourselves again of those basics. Here are some of them:

1. Any number name has very many acceptable substitutions, because the set corresponding to that number can be seen in many ways as a combination of subsets.
    Example: the number whose standard name is 9 can also be expressed as $3 + 4 + 2, 2 + 6 + 1, 2 + 1 + 6, 2 + 7$, etc.

2. Every subtraction sentence is linked with an addition sentence. This is because when a subset is removed from a set, the original situation is restored when the same subset is put back.
    Example: If 4 objects are removed from 9, 5 objects are left. If the 4 objects are added to the 5 the 9 objects are once more seen.
    i.e.,
    If 9 - 4 = 5 then 5 + 4 = 9
    If 5 + 4 = 9 then 9 - 4 = 5

This is true for any three numbers which can be substituted for the 9, 4 and 5 and give a true sentence.

3.   If to a set of 9 objects are added 2 more objects and then as previously (No. 2 above) 4 objects are removed, what is left this time is the same as before *increased by the 2.*
i.e.,
$$\text{If } 9-4 = 5 \text{ then } (9 + 2)-4 = 5 + 2$$
Similarly,
$$(9 + 3)-4 = 5 + 3$$
$$(9 + 4)-4 = 5 + 4$$
$$(9 + 5)-4 = 5 + 5$$
$$(9 + 100)-4 = 5 + 100 \ldots \text{ and so on.}$$

The pattern is followed for any three numbers substituted for the 9, 4 and 5 which give a true sentence, and with any number substituted for the two instances of 100 in the last line above.

Or, in the jargon used with colored rods,
$$\text{blue - pink } = \text{ yellow}$$
compels us also to accept that,
$$(\text{blue} + \text{red})-\text{pink} = \text{yellow} + \text{red}$$
$$(\text{blue} + \text{green})-\text{pink} = \text{yellow} + \text{green, etc.}$$

4.   If from a set of 9 objects we remove three objects and *then* remove the 4 objects as previously, what is left is the same as before *decreased* by the 3:
i.e. If $9-4 = 5$ then $(9-3)-4 = 5-3$

or $9-3-4 = 5-3$

and $(9-4)-3 = 5-3$

or $9-4-3 = 5-3$

Substituting for the 3 we could also write:

$(9-2)-4 = 5-2$

$(9-1)-4 = 5-1$ etc.

The pattern is also followed for any three numbers substituted for the 9, 4 and 5 which give a true sentence and with any number substituted for the two instances of 1 in $(9-1)-4 = 5-1$, so long as the substituted number can be subtracted from 9 and from 5. For we can apply this to counters and be satisfied, and in the case of rods:

Note: Maybe it seems that 6 or 7 cannot be substituted in the sentences above, to give

$$(9 - 6) - 4 = 5 - 6$$
$$(9 - 7) - 4 = 5 - 7$$

Nevertheless, it *would* be convenient to be able to write this. Such examples can hardly apply to taking away counters because we cannot remove 7 counters from a set of five. With rods, however, we *do* get an interpretation, 'five minus seven' perhaps being given a local class standard name of 'come back two', 'two below zero' or the conventional and traditional 'negative two'.

5. To summarize other basic principles rather than write about them at length:
    a) If $9 - 4 = 5$
       then $(9 + 3) - (4 + 3) = 5$
               $(9 + 4) - (4 + 4) = 5$
               $(9 + 5) - (4 + 5) = 5$
               $(9 + 6) - (4 + 6) = 5$
            $(9 + 100) - (4 + 100) = 5$ . . . and so on
    b) If $9 - 4 = 5$
       then, $(9 - 1) - (4 - 1) = 5$
                $(9 - 2) - (4 - 2) = 5$
                $(9 - 3) - (4 - 3) = 5$ . . . and so on

       and possibly,
                $(9 - 5) - (4 - 5) = 5$
                $(9 - 6) - (4 - 6) = 5$ . . . and so on, although we
                                          cannot show this with
                                          objects.
    c) If $9 - 4 = 5$
       then $9 - (4 + 1) = 5 - 1$
                     $9 - (4 + 2) = 5 - 2$
                     $9 - (4 + 3) = 5 - 3$ . . . and so on

"Oh!", someone could complain, "All this is so tedious to wade through and I am no mathematician." True, but if it depended on having to understand written explanation very few people would bother. Fortunately, it is all extremely simple when it is done with objects, although the actions needed cannot be recorded simply in print. All the written words can do is to invite the reader to try the actions with small objects like counters, stones or rods. For greater details references may be made to the Bibliography.

With the basic principles mastered or at least with students having available counters or rods as referents to check the legitimacies of principles, any subtraction pair can be transformed into other, equivalent, pairs and thence to its corresponding standard.

> e.g., $531 - 174$ can be transformed to $431 - 74$ (by subtracting 100 from each number – Principle 5b)
> then to $401 - 44$ (Principle 5b)
> then perhaps, to $391 - 34$
>   to $361 - 4$
>   to 357

Alternatives:

$$531 - 174 \rightarrow 537 - 180 \rightarrow 557 - 200 \rightarrow 357$$
$$331 + 6 + 20 = 357 \text{ (Pple 2)}$$

$$\begin{array}{c} 531 \\ -174 \end{array} \rightarrow \begin{array}{c} 531 \\ -180 \end{array} \rightarrow \begin{array}{c} 531 \\ \underline{200} \end{array} \rightarrow 331$$

$$\begin{array}{c} 531 \\ -174 \end{array} \rightarrow \begin{array}{cc} 500 & + 31 \\ \underline{170} & \underline{+\ 4} \end{array} \rightarrow 330 + 27 \rightarrow 357 \text{(Pple 1)}$$

This last transformation is in accord with some text strategies which change the 'borrowing' technique to one called 'regrouping'. That means using alternate non - standard names, choosing those which enable the student to subtract more easily.

Another regrouping possibility is:

$$\begin{array}{c} 531 \\ -174 \end{array} \rightarrow \begin{array}{c} 520 + 11 \\ \underline{170 + \ 4} \end{array} \rightarrow \begin{array}{c} 520 \\ \underline{170} \end{array} + 7 \rightarrow \begin{array}{c} 550 \\ \underline{200} \end{array} + 7 \rightarrow 350 + 7 \rightarrow 357$$

It is *not* implied that the steps here have to be written. The one who is computing chooses the steps he wants and uses writing when it helps *him*.

Discussion in class between students will lead to wider recognition of the many alternatives possible, all founded on the same basic principles. As experience grows, each student will begin to select techniques more economic for himself, according to his own unique insights and needs. Which particular technique is used will depend only upon the actual number names in question, the student's insight and strategy suggested at the moment of deciding to process and perhaps specific considerations suggested by the teacher.

Such processes can rightly be called the 'algebra of subtraction', for the actual number names used are irrelevant to the utilization of the principles basic for *all* cases. The aim is to master that algebra for every possibly pair of numbers presented, rather than use gimmicks based mostly upon memory.

A variation of the application of one basic principle is worth noting because it is impressive to those who believe that subtraction is hard. Yet anyone can succeed in it. It capitalizes on the principle that an equivalent subtraction pair is obtained by increasing each number by the same. Originally introduced by Gattegno it can be called 'speedometer subtraction'.

Speedometer Subtraction

Imagine two cars travelling alongside each other and suppose the odometers show 53104 on car A, 47865 on car B. In considering the standard name for the difference of these numbers, 53104 – 47865, we can say that when the cars have gone another 4 miles (or kilometers or any other unit), the difference remains the same and the equivalent pair is 53108, 47869.

						A:	53138	B:	47899
In another	30	miles we have,							
"	"	100	"	"	"		53238		47999
"	"	2000	"	"	"		55238		49999
"	"	50000	"	"	"		105238		99999
"	"	1	"	"	"		105239		100000

The standard name is clearly 5239.

A verbal patter can, if wished be added:

Looking at 53104 and 47865 we ask, "What is the complement of 5 in 9? (5 is the units digit of 47865)." The answer being 4, this is added to 53104 giving 53108. We now have 53108 and 47869. "What is the complement of 6 in 9? (47869)". The answer is 3, so 3 is added to 0 in 53108, giving 53138, and so on, until every digit of the second number is 9.

Most people, children and adults, have become excited in cars when they see the mileage reading "all 9's" and watching when, one mile further on, all the 0's come up instead, the 1 being imagined on the left of the figures shown.

In the calculation, however, all we do is to add 1 to the first number and drop the 1 from the first left hand position.

$$\frac{105238}{99999} \rightarrow \frac{105239}{100000} \rightarrow 5239$$

If one prefers, one can use 'the complement in 10' in the units position rather than adding the 1 later.

A slight but expected variation occurs if the upper digit is greater than its corresponding lower partner.

$$\text{e.g.} \quad \frac{53194}{-47865} \rightarrow \frac{53198}{-47869}$$

In the 'tens place' the 'complement of 6 in 9 is 3' so 3 has to be added to the upper 9, making 12. Naturally the ten has to be carried:

$$\frac{53198}{-47869} \rightarrow \frac{53228}{-47899} \quad \text{... and so on.}$$

Finally, after practise, the 'answer' is written for any given example.

$$
\begin{array}{r}
7213486 \\
-2541625 \\
\hline
10 \\
15 \\
117 \\
1466 \\
\hline
14671860 \\
\end{array}
\quad +1 \text{ 'more mile', less } 10{,}000{,}000
$$

Answer: 4671861

On other planets

On 'other planets' the speedometer subtraction lends itself well. Instead of using 9's we shall have, of course, to use the local standard for 10 - 1.

On the 'pink planet':
$$3 + 1 = 10$$
$$33 + 1 = 100$$
$$333 + 1 = 1000 \ldots \text{and so on.}$$

$$
\begin{array}{cccc}
231021 & 231022 & 1101122 & 1101123 \\
-123232 & -123233 & -333333 & -1000000 \\
\hline
& & & 101123
\end{array}
$$

$\rightarrow$ $\rightarrow$ $\rightarrow$

Whether this in all contexts is the best strategy is *not* the point. It *is* legitimate and especially interesting when large numbers are involved. Intermediate school children find it compelling.

8

The Multiplication Table Revisited

As soon as a sum contains a repetition of one of its components multiplication can be said to arise. A substitute for $7 + 3 + 3$ is $7 + 2 \times 3$, as the 3 occurs twice in the sum. 2×3 is read 'two times three', although sometimes 2×3 is interpreted as 'two multiplied by three'. The associated sum in the latter case is $2 + 2 + 2$.

2×3 is called a *product*. 6 is its standard name. If 2×3 is read as '2 times 3' the 2 is called the operator because it qualifies the 3; if '2 multiplied by 3' then 3 is the operator. Although there are these two ways of reading 2×3 the ambiguity is usually not confusing because $2 \times 3 = 3 \times 2$.

Thus, multiplication receives a working definition of 'repeated addition', though it can also be derived from a consideration of sets. Once a product is accepted as equivalent to certain sums other products can be substituted.

$$\begin{aligned}
\text{e.g. } 4 + 4 + 4 + 4 + 4 &= 5 \times 4 \\
&= (4 \times 4) + (1 \times 4) \\
&= (3 \times 4) + (2 \times 4) \\
&= (2 \times 4) + (3 \times 4) \\
&= (2 \times 4) + (2 \times 4) + (1 \times 4) \\
&\qquad \ldots \text{and so on.}
\end{aligned}$$

For such practice the standard names of all the separate products are not needed, only the awareness of what later may be called the 'distributive principle', or perhaps here the 'undistributive principle' because the former is usually applicable to a product like $(2 + 2 + 1) \times 4$. For beginners the above substitutions come either from inspection of the number of components repeated in a sum, the number of equal size piles of counters or the number of rods of one color in a train of rods.

As with addition (chapter 6) a table can now be constructed, with products as entries. The headings are normally 0–10, or 1–10, but many alternatives are possible.

x	0	1	2	3	4	5	6	7	8	9	10
0	0x0	0x1	0x2	0x3	0x4	0x5	0x6	0x7	0x8	0x9	0x10
1	1x0	1x1	1x2	1x3	1x4	1x5	1x6	1x7	1x8	1x9	1x10
2	2x0	2x1	2x2	2x3	2x4	2x5	2x6	2x7	2x8	2x9	2x10
3	3x0	3x1	3x2	3x3	3x4	3x5	3x6	3x7	3x8	3x9	3x10
4	4x0	4x1	4x2	4x3	4x4	4x5	4x6	4x7	4x8	4x9	4x10
5	5x0	5x1	5x2	5x3	5x4	5x5	5x6	5x7	5x8	5x9	5x10
6	6x0	6x1	6x2	6x3	6x4	6x5	6x6	6x7	6x8	6x9	6x10
7	7x0	7x1	7x2	7x3	7x4	7x5	7x6	7x7	7x8	7x9	7x10
8	8x0	8x1	8x2	8x3	8x4	8x5	8x6	8x7	8x8	8x9	8x10
9	9x0	9x1	9x2	9x3	9x4	9x5	9x6	9x7	9x8	9x9	9x10
10	10x0	10x1	10x2	10x3	10x4	10x5	10x6	10x7	10x8	10x9	10x10

When standard names are known for some of the products a second table can be created. A student knows some of them by the time he meets the multiplication table. If not, he can use a strategy of changing a product to its corresponding sum or alternatively get standard names by using concrete objects.

The completed table is that traditionally used and considered as a most important piece of learning in arithmetic. That is still true although we are going to study much more than in previous years.

x	0	1	2	3	4	5	6	7	8	9	10
1	0	1	2	3	4	5	6	7	8	9	10
2	0	2	4	6	8	10	12	14	16	18	20
3	0	3	6	9	12	15	18	21	24	27	30
4	0	4	8	12	16	20	24	28	32	36	40
5	0	5	10	15	20	25	30	35	40	45	50
6	0	6	12	18	24	30	36	42	48	54	60
7	0	7	14	21	28	35	42	49	56	63	70
8	0	8	16	24	32	40	48	56	64	72	80
9	0	9	18	27	36	45	54	63	72	81	90
10	0	10	20	30	40	50	60	70	80	90	100

Because in chapter 6 we listed at considerable length many patterns of the addition table we shall not do so in as like detail with the multiplication table, but suggest some of the questions which can be asked to stimulate a pattern search. The list of suggestions this time corresponds closely to that previously but this must not be taken to imply that either order of patterns is the best or even a good one. It is certainly not a likely order for any individual involved in a learning situation more flexible than following a book.

To this end a reader is encouraged *to pause at this point* and find patterns in the finished multiplication table for himself. Perhaps a study like that may take more than one sitting, over a few days, before reference is made to the list of exercises which corresponds to the order and contents of the list in chapter 6.

Exercise A

What alternative headings are there for a multiplication table? Devise some, writing in the entries both in product form and as standard names.

x	1	2	3	4	5	6	7	8	9	10	11
1	1	2	3	4	5	6	7	8	9	10	11
2	2	4	6	8	10	12	14	16	18	20	22

Exercise B

Study rows and columns. How many standard names for products are needed to complete the table?

Exercise C

Study entries in NE - SW diagonals. Read each in two directions.

Exercise D

In the addition table each row was very alike the previous row and the one following. Is this true for the multiplication table?

Exercise E

Choose a column and inspect the sequence of the right hand digits. Is there another column with the same digits, perhaps in a different order?

What pairs of columns have the same subset of digits?

x	0	1	2	3	4	5	6	7	8	9	10
0			0						0		
1			2						8		
2			4						16		
3			6						24		
4			8						32		

5	10	40
6	12	48
7	14	56
8	16	64
9	18	72
10	20	80
		↑

Exercise F

The numbers in the main NW - SE diagonal are 0, 1, 4, 9, 16 . . . What are their corresponding products? These are called 'square numbers'. Are there corresponding square shapes, each with one of the square numbers at the SE corner?

Exercise G

Study the differences between pairs of neighboring numbers in each of the NW - SE diagonals.

```
        e.g.          0   1   4   9   16   25 . . .
first differences:      1   3   5   7    9      . . .
second differences:       2   2   2   2          . . .
```

Example:

x	0	1	2	3	4	5	6	7	8	9
0		0	*2*							
1			2	*4*						
2				6	*6*					
3					12	*8*				
4						20	*10*			
5							30	*12*		
6								42	*14*	
7									56	*16*
8										72
9										

On a copy of the multiplication table write such first differences for some of the NW - SE diagonals in position between the entries. When about a dozen of the differences are written, copy them in the same positions onto blank paper.
Does this not give part of the addition table?

Should this be not surprising?
Insert appropriate headings and compare with the second table in chapter 6.

Exercise H

A multiplication table can also be seen to be a division table.
For example, the standard for $24 \div 3$ is located by tracking 24 in column 3.   24 is in row 8.
Therefore, $24 \div 3 = 8$ (because $8 \times 3 = 24$)

Are there exceptions?
Try, $9 \div 0$;   there is no 9 in column 0
or, $4 \div 0$;   there is no 4 in column 0.
We say $9 \div 0$ has *no meaning*. There is no number n such that $n \times 0 = 9$.
Try also $0 \div 0$. There are plenty of 0's in column 0.

One of them is in row 0.
So apparently $0 \div 0 = 0$.
Another 0 is in row 1, another in row 2
... So apparently $0 \div 0 = 1, 0 \div 0 = 2 \ldots$

This is silly, for we cannot have $0 \div 0$ as an equivalent for 0, for 1, 2, 3 ... and so on. Therefore, $0 \div 0$ has *no meaning*.

Division by 0 is *meaningless*.

Exercise I

Choose any entry in the multiplication table. For example, begin at the entry 2 in column 2. If a move is made 1 place East followed by another 2 steps South, what entry is reached?

x	0	1	2	3	4	5	6	7	8	9	10
0	0	0	0	0	0	0	0	0	0	0	0
1	0	1	2 →3	4	5	6	7	8	9	10
2	0	2	4	6	8	10	12	14	16	18	20
3	0	3	6	9 →12	15	18	21	24	27	30
4	0	4	8	12	16	20	24	28	32	36	40
5	0	5	10	15	20 →25	30	35	40	45	50
6	0	6	12	18	24	30	36	42	48	54	60
7	0	7	14	21	28	35 →42	49	56	63	70
8	0	8	16	24	32	40	48	56	64	72	80
9	0	9	18	27	36	45	54 →63	72	81	90
10	0	10	20	30	40	50	60	70	80	90	100

If these combined moves are coded as (1E, 2S), give the sequence of numbers reached, beginning with the 2.   (2, 9, 20 ...)

What are the first and second differences of these numbers?

If the same moves transform any other number in the table and successively the numbers after than, what can be noticed about all second differences?

Play the same game using moves: (1E, 2S), (1E, 4S), (2E, 3S) ... and so on. Can second differences in each case be decided by using a pattern from the code of each move, from (1E, 3S), (2E, 3S) ... and so on?

Exercise J

x	0	1	2	3	4	5	6	7	8	9	10
0	0	0	0	0	0	0	0	0	0	0	0
1	0	1	2	3	4	5	6	7	8	9	10
2	0	2	4	6	8	10	12	14	16	18	20
3	0	3	6	9	12	15	18	21	24	27	30
4	0	4	8	12	16	20	24	28	32	36	40
5	0	5	10	15	20	25	30	35	40	45	50
6	0	6	12	18	24	30	36	42	48	54	60
7	0	7	14	21	28	35	42	49	56	63	70
8	0	8	16	24	32	40	48	56	64	72	80
9	0	9	18	27	36	45	54	63	72	81	90
10	0	10	20	30	40	50	60	70	80	90	100

Study the two products formed by the entries in opposite diagonal corners of *any* rectangle drawn on the multiplication table. What is the relationship between every pair of these products?

Exercise K

The tapping game in which pairs of entries were selected as belonging to a family can be used with the multiplication table, just as previously with the addition table. For details of this reference can be made forward to chapter 12.

As an exercise here, however, readers are invited to study what a similarly patterned sequence of 'pairs of taps' suggests when the multiplication table is substituted for the addition table.

x	0	1	2	3	4	5	6	7	8	9	10
0	0	0	0	0	0	0	0	0	0	0	0
1	0	1	2	3	4	5	6	7	8	9	10
2	0	2	4	5	8	10	12	14	16	18	20
3	0	3	6	9	12	15	18	21	24	27	30

| 4 | 0 | 4 | 8 | 12 | (16) | 20 | 24 | (28) | 32 | 36 | 40 |
| 5 | 0 | 5 | 10 | 15 | [20] | 25 | 30 | [35] | 40 | 45 | 50 |

$\downarrow$ $\downarrow$

Exercise L

The second tapping game previously applied to the addition table (L in chapter 6) can also be tried on the multiplication table.
Readers are challenged to explore the consequences of the pattern in which 3 pairs of entries in one row are chosen.
What is the traditional and conventional form of writing such triplets of pairs?
Details are given in chapter 13.

Exercise M

$- 16$
$- 12$
$- 8$
$- 4$

$- 16$ $- 12$ $- 8$ $- 4$

x	0	1	2	3	4	5	6	7
0	0	0	0	0	0	0	0	0
1	0	1	2	3	4	5	6	7
2	0	2	4	6	8	10	12	14
3	0	3	6	9	12	15	18	21
4	0	4	8	12	16	20	24	28
5	0	5	10	15	20	25	30	35
6	0	6	12	18	24	30	36	42
7	0	7	14	21	28	35	42	49

As previously with the addition table, the entries in all columns and rows can be extended beyond 0. Column 4, for example, shows a sequence ... 12, 8, 4, 0 each number being 4 less than the previous. If there were to be a next entry we would expect it to be some representation of 0 - 4, the next 0 - 4 - 4, 0 - 8, or 0 - 2 x 4 ... and so on. The standard forms of these are $- 4$, $- 8$, $- 12$ ... and so on.
Extend each column and row to provide entries in the NE, SW and NW portions of the multiplication table. The headings, too, will need extended entries.

	-5	-4	-3	-2	-1	x	0	1	2	3	4	5	6
...	25	20	15	10	5	- 5	0	- 5	- 10	- 15	- 20	- 25	- 30
...	20	16	12	8	4	- 4	0	- 4	- 8	- 12	- 16	- 20	- 24
...	15	12	9	6	3	- 3	0	- 3	- 6	- 9	- 12	- 15	- 18
...	10	8	6	4	2	- 2	0	- 2	- 4	- 6	- 8	- 10	- 12
...	5	4	3	2	1	- 1	0	- 1	- 2	- 3	- 4	- 5	- 6
...	- 5	- 4	- 3	- 2	- 1	x	0	1	2	3	4	5	6
...	0	0	0	0	0	0	0	0	0	0	0	0	0
...	- 5	- 4	- 3	- 2	- 1	1	0	1	2	3	4	5	6
...	- 10	- 8	- 6	- 4	- 2	2	0	2	4	6	8	10	12
...	- 15	- 12	- 9	- 6	- 3	3	0	3	6	9	12	15	18
...	- 20	- 16	- 12	- 8	- 4	4	0	4	8	12	16	20	24
...	- 25	- 20	- 15	- 10	- 5	5	0	5	10	15	20	25	30
...	- 30	- 24	- 18	- 12	- 6	6	0	6	12	18	24	30	36

Exercise N

Extract many products from the extended table until it can be done correctly without resorting to the written entries.

$$\text{e.g. } ^-2 \times 3 = \, ^-6$$

If wished the non - negative numbers can be termed 'positive', written with a $+$ sign slightly raised: $^-2 \text{ x } ^+3 = \, ^- 6$, though conventionally the $+$ sign is often omitted in contexts which make it clear that both negative and positive numbers are being used. The negative and positive numbers, including zero, are called *integers*.

Exercise O

Study the applicability of all patterns of the unextended multiplication table, now that negative numbers are available and all rows and columns have theoretically an endless number of entries.
Do the previous patterns continue unchanged in all the extensions?

Exercise P₁

The multiplication table used so far in this study of patterns has had number names in the common base. Previously, however, in chapter 3, an excursion was made to the 'pink planet' so we can, if we wish, study the multiplication in *that* base. (Base IV, if we prefer this name rather than 'pink planet')
Check that this is correct:

x	1	2	3	10
1	1	2	3	10
2	2	10	12	20
3	3	12	21	30
10	10	20	30	100

Study these patterns in the above table:
- Entries in the NE - SW diagonals
- Righthand digits of names in columns
- Square numbers
- Differences of neighboring pairs in every NW - SE diagonal
- Division sentences
- Products of opposite corner entries of any rectangle
- Extensions in N, S, E, W, directions, to negative entries.

Exercise P $_2$

Repeat the procedures of Ex. P $_1$ on a multiplication table of any other base: Base IX, VIII, VII . . . and so on.

Exercise Q

The Vedic transformation can be used on the multiplication table (Base X).
As previously (chapter 6), 13 will be replaced by 1 + 3 or 4
89 will be replaced by 8 + 9 or 17, and
that by 1 + 7 or 8.
This provides another table in which patterns can easily be seen.
Examine: repetitions, columns and rows, diagonals, symmetries, the locations in the table for each digit in turn.

x	1	2	3	4	5	6	7	8	9
1	1	2	3	4	5	6	7	8	9
2	2	4	6	8	1	3	5	7	9
3	3	6	9	3	6	9	3	6	9
4	4	8	3	7	2	6	1	5	9
5	5	1	6	2	7	3	8	4	9
6	6	3	9	6	3	9	6	3	9
7	7	5	3	1	8	6	4	2	9
8	8	7	6	5	4	3	2	1	9
9	9	9	9	9	9	9	9	9	9

Exercise R

The last pattern mentioned here is one which applies only to the multiplication table.
Each entry can be changed in this way, for example:

$$11 \div 5 = 2 \text{ and } 1 \text{ remainder}$$
$$12 \div 5 = 2 \text{ and } 2 \text{ remainder}$$
$$13 \div 5 = 2 \text{ and } 3 \text{ remainder}$$
$$14 \div 5 = 2 \text{ and } 4 \text{ remainder}$$
$$15 \div 5 = 3 \text{ and } 0 \text{ remainder}$$
$$16 \div 5 = 3 \text{ and } 1 \text{ remainder} \ldots \text{ and so on.}$$

Note that 11 and 16, divided by 5, give the same remainder of 1.

For the entries 11, 16 in the multiplication table, 1 can consequently be substituted. Similarly for each entry, the remainder, when the entry is divided by 5, can be written instead. A new table is thus formed.

x	1	2	3	4	5	6	7	8	9	10
1	1	2	3	4	0	1	2	3	4	0
2	2	4	1	3	0	2	4	1	3	0
3	3	1	4	2	0	3	1	4	2	0
4	4	3	2	1	0	4	3	2	1	0

Study patterns in this 'divisor 5, remainder' table.
Form also tables: 'divisor 6, remainder', 'divisor 7, remainder' . . . and so on.
Compare each table with the others.

Thus, the multiplication table is seen to be much more than an arrangement of standard names corresponding to products, to be memorized. It can be studied as a rich source of patterns and as a point of regular departure giving leads to many other arithmetical and mathematical topics.

X	1	2	3	4	5	6	7	8	9	10	11	12
1	1	2	3	4	5	6	7	8	9	10	11	12
2	2	2x2	6	2x4	15	12	2x7	19	2x9	20	22	2x12
3	3	3x2	3x3	3x4	3x5	3x6	3x7	3x8	3x9	3x10	33	3x12
4	4	4x2	4x3	4x4	4x5	4x6	4x7	4x8	4x9	40	44	4x12
5	5	5x2	15	20	25	5x6	5x7	5x8	5x9	5x10	55	5x12
6	6	6x2	6x3	6x4	6x5	6x6	6x7	6x8	6x9	6x10	6x11	6x12
7	7x1	7x2	7x3	7x4	7x5	7x6	7x7	7x8	7x9	7x10	77	7x12
8	8x1	8x2	8x3	8x4	8x5	8x6	8x7	8x8	8x9	8x10	88	8x12
9	9	9x2	9x3	9x4	9x5	9x6	9x7	9x8	9x9	9x10	9x11	9x12
10	10x1	10x2	10x3	10x4	10x5	10x6	10x7	10x8	10x9	10x10	10x11	10x
11	11x1	11x2	11x3	11x4	11x5	11x6	11x7	11x8	11x9	11x10	11x11	11
12	12	12x2	12x3	12x4	12x5	12x6	12x7	12x8	12x9	12x10	12x11	

Times table conversion to standard names

9

Factors and Multiples
An Alternative Look at Previous Work

When there has been an entry into product names and their standard equivalents, factors and multiples are hardly difficult new ideas or skills. Yet in school they are frequently sources of difficulty, probably because the names 'factor' and 'multiple' do not become familiar enough in interesting and challenging ways before they are used in fraction work.

Factors from the Times Table

A beginning can be made by saying that 7 is a factor of 7 x 6. So is 6. Are there others? Since other equivalent products are 1 x 42, 2 x 21, 3 x 14, we can say that 1, 2, 3, 6, 7, 14, 21, and 42 are all factors of 42. If the multiplication table is known in the chart form, then factors can be extracted. Whenever 20, for example, appears as an entry its equivalent product name reveals factors, or reference can be made to the headings of the rows and columns in which 20 lie.

Entries in any row (or column) give the multiples of the heading of that row (or column) with the exception of 0. One could say that 0 is a multiple of 20, say, in the sense that 0 = 0 x 20, but then 0 would be a multiple of *every* number! It is generally accepted that 0 is not included as a multiple of any number.

Is 0 a *factor* of 20, or of any other number? No, because 20 cannot be found in any row (or column) headed 0. Or put in a different way, there is no number which with 0 gives a product equivalent to 20.

Factors with Rods

If rods are used, an entry to factors can occur when trains equivalent to the length of any other train are made. Red is a factor of orange, for example, since there exists an all - red train equal in length to the orange rod. White is also a

106

factor, light green is not, nor is pink; yellow is a factor; dark green, black, brown, blue are not. However, orange *is* a factor of orange, of 'itself'.

From such initial work children can be asked, "What are the factors of ten?" If, by that time, they are quite familiar that with white as the unit, the orange is 10, they will readily provide 1, 2, 5, and 10 as the factors.

Factors with Counters

Sets represented by counters or pebbles give a slightly more cumbersome introduction but, of course, still a valid one. With a pile of counters in front of him, a student is asked, "What sets, each of equivalent size to every other, combined, would give *this* pile?" or more technically, perhaps, "What equivalent sets in union would give this set?" Certainly it will always be possible to see each set in single elements – their union *is* the given set. So a single element subset is a factor of every set. Further than that, however, the factors will depend on the size of the given set. Although that is also true no matter what materials are used, the passage from one rod which does not yield many factors to another which is rich in factors is perhaps a more satisfying activity than the need to count the pebbles in each pile.

Factors of Factors

Children can thus find factors of various numbers, the teacher suggesting they begin perhaps with 8, 10, 12, or 9, to be followed by the smaller and larger numbers. At first some factors for each number can be recorded but as the study develops *all* the factors for each number can be sought. A cooperative enterprise by the whole class can, in a few hours, produce a large display on which factors of various numbers have been recorded:

No.	Factors	No.	Factors	No.	Factors
8	1,2,4,8	4	1,2,4	20	1,2,4,5,10,20
10	1,2,5,10	3	1,3	24	
12	1,2,3,4,6,12	2	1,2	30	. . .and so on, any
9	1,3,9	1	1 (maybe 1,1)	36	choices of numbers
7	1,7	11	1,11	48	being made
6	1,2,3,6	13	1,13	60	
5	1,5	14	1,2,7,14	100	

This provides a plateau on which one can pause temporarily before taking another leap forward. Data has been gathered, discussed, checked, probably amended. The results can, therefore, be looked at with confidence and patterns gleaned from them. Students themselves may well put questions like the following; otherwise the teacher can ask them:

Does every number have factors?
Does zero have a factor?
Which numbers have the least number of factors?
Which have just one factor?
Which have just two factors? (These are called *prime* numbers.)
Which numbers have more than two factors? (These are called *composite* .)
Which numbers have 2 as a factor? (These are called *even* numbers.)
Which numbers have the number itself as one of its factors?
How many factors has each number?
Do the larger numbers have more factors than the smaller?
Which numbers have an even number of factors, which odd (or non - even) . . . and so on?
Do factors arise in pairs, such that the product of the two factors of a pair are equivalent to the number? (One factor can be termed the 'complement' of the other. For addition, 2 is the complement of 8, in 10. For multiplication 2 is the complement of 5, in 10.)
If one number is a factor of another, are all the factors of the first also factors of the second? Is the converse true? – that is, are all the factors of the second number also factors of the first?
How many multiples does a number have?
If one number is a multiple of a second number, and the second number is a multiple of a third, is the given number always a multiple of the third? Give examples.
Are there numbers which have the same factors? (They are called 'common factors'.)
Are there numbers which have the same multiples? (Such are called 'common multiples'.)
Of all the common factors which two or more numbers may share, is one of the common factors the least? (It is called the 'least common factor' or the 'lowest common factor'.) Is there also a 'greatest or highest common factor'?
Is there of several numbers a 'lowest common multiple' and a 'highest common multiple'? Find some by inspection of a factors and multiples chart.

Crosses and Towers with Rods

If students are familiar with the strategy of making crosses and towers with the rods, another set of activities with factors and multiples is available. The tower (red x light - green x black), for example, corresponds to the product 2 x 3 x 7. It is a multiple of 2 since the tower is equivalent to a train of 21 red rods; a multiple of 3 because the same length train is equivalent to 14 light - green rods and so on. The complete set of factors is in fact, (2, 3, 7, 2 x 3, 2 x 7, 3 x 7, 1, 2 x 3 x 7) every member of which is seen in the tower except possibly the 1. But 1 is a factor of every number!

The multiples of 2 x 3 x 7 can be shown by placing another rod across the top, giving 2 x (2 x 3 x 7), 3 x (2 x 3 x 7), 4 x (2 x 3 x 7) . . . and so on, ad infinitum!

Common Factors and Multiples

> Red is a factor of orange ⎤   So red is a factor *common*
> Red is a factor of brown ⎦   to orange and brown.
>   (strictly, of course, red implies 'red *length*')
> 2 is a factor of 10 ⎤ So 2 is a factor common to 10 and 8
> 2 is a factor of 8 ⎦

A common factor of 8, 12 and 20 is 2. Another is 4; another 1. Of these, 4 is the greatest or highest. Therefore, the highest common factor (H.C.F.) of 8, 12 and 20, is 4. The lowest common factor is 1, but 1 is always the lowest common factor of any two or more numbers. Once we see this it is hardly worth stressing again!

> Multiples of 8 are 8, 16, 24, 32 . . . A row in the multiplication table
> Multiples of 12 are 12, 24, 36, 48 . . . also a row in the multiplication table
> Multiples of 20 are 20, 40, 60, 80 . . . also a row in the multiplication table

Is there a number which occurs in every one of these three rows, even if it has not been written so far? Yes, 120 does. So do 240, 360, 480, and so on. All these are common multiples of 8, 12, and 20; 120 is the least of them. If they are not yet recorded in standard form then we may have 8 x 15, 12 x 10 and 20 x 6. They are equivalent because

$$8 \times 15 = 2 \times 2 \times 2 \times 3 \times 5$$
$$12 \times 10 = 2 \times 2 \times 3 \times 2 \times 5$$
$$20 \times 6 = 2 \times 2 \times 5 \times 2 \times 3$$

Common Multiples

Common multiples can easily be shown with rods. One for red and light - green rods, for instance, arises by constructing two trains alongside each other, one entirely of red rods, one of light - green rods. On inspection we see that after a while the red and green trains are of the same length. This indicates a common multiple. More common lengths, but of increasing length, occur again and again regularly.

The first common lengths of red and light - green trains occur at the length of 3 red or 2 light - greens. The next is at 6 reds or 4 light - greens; the next 9 reds or 6 light - greens . . . and so on. The shortest of these is the first, given by 2 x 3 or 3 x 2, the products of 2 and 3.

In some cases inspected, however, the lowest common multiple is *not* the product of the numbers. For 4 and 8, for example, the lowest common multiple is 8 not 32, though 32 is certainly one of the common multiples. For 10 and 15, 150 is certainly a common multiple, but 30 is also and it is the least! How shall we know when the product gives the least common multiple and when not? Rod trains provide the awareness.

If we consider the common multiples of 10 and 15, we can inspect an orange train and an orange - plus - yellow train. But this is equivalent to inspecting one train comprising repetitions of 2 yellow and another of repetitions of 3 yellows. The lowest common multiple of 2 yellows and 3 yellows is 2 x 3 yellows. So the lowest common multiple of 10 and 15 is 2 x 3 x 5 and *not* 10 x 15.

Another example will probably confirm the pattern. For 6 and 10 the
equivalent lengths are 3

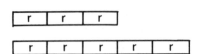

reds and 5 reds. Their shortest common multiple is 3 x 5 reds. Thus the lowest common multiple of 6 and 10 is 3 x 5 x 2 = 30 and not 6 x 10.

The general pattern is that the product of the given numbers certainly gives a common multiple, but to get the lowest of all the common multiples the product is divided by the common factors.

If the two numbers are represented by a and b, and a and b are prime to each other – that is, they have no common factor (other than 1 of course) – then the lowest common multiple is a x b.

If the two numbers are a x b, c x b, and b is the highest common factor, then a common multiple is certainly a x b x c x b. But a x b x c is also a common multiple, is less than the product of the two given numbers and is the lowest common multiple.

> Should there be 3 numbers, a, b, c, the lowest common multiple is a x b x c;
> for a x n, b x n, c x n, the lowest common multiple is a x b x c x n, or the product divided by n²;
> for a x n, b x n, c x b, the lowest common multiple is a x b x n x c or the product divided by b x n . . . and so on.

Primes

A study of prime numbers can include the commonly called 'Sieve of Erastothenes', though knowing the strategy by that title is surely not necessary. This amounts to a chart displaying the whole numbers from 1 to 100 or further. All numbers which are composite – *not* primes – are then crossed out. Perhaps one begins with 10, 20, 30 . . . 100, followed by 4, 8, 12, 16 . . . Gradually only the primes are left. The composite numbers, as it were, have slipped through as do the small particles of a sieve. The prime numbers remaining are: 1, 2, 3, 5, 7, 11, 13, 17, 19, 23, 29, 31,

1 2 3 4 5 6 7 8 9 10

11 12 13 14 15 16 17 18 19 20

21 22 23 24 25 26 27 28 29 30

31 32 33 34 35 36 37 38 39 40

41 42 43 44 45 46 47 48 49 50

51 52 53 54 55 56 57 58 59 60

61 62 63 64 65 66 67 68 69 70

71 72 73 74 75 76 77 78 79 80

81 82 83 84 85 86 87 88 89 90

91 92 93 94 95 96 97 98 99 100

37, 41, 43, 47, 53, 59, 61,
67, 71, 73, 79, 83, 89, 97.

There is, however, little value in remembering precisely which numbers up to 100 are primes though there are patterns of composites worth spotting. In the common base 10, multiples of 2 always end in 0, 2, 4, 6, or 8. Multiples of 4 and 8 do too. These patterns probably arose from study of the righthand digits in the columns of the multiplication table. Multiples of 10 end in 0 and multiples of 5 alternate with 0 and 5 for the unit digit.

Multiples of 3 have a pattern, too. Consider a standard name written AB. It is equivalent to $(10 \times A) + B$. "A - tyB", and *that* is equivalent to $(9 \times A) + A + B$. $9 \times A$ is a multiple of 3, and incidentally of 9, so if $A + B$ is a multiple of 3 or 9, then the number AB is so, too.

The rules change somewhat, of course, on other planets. To cite one example, consider the 'pink planet'. The multiplication table is

x	1	2	3	10
1	1	2	3	10
2	2	10	12	20
3	3	12	21	30
10	10	20	30	100

Multiples of 2 have 0 or 2 as their righthand digit.
Multiples of 10 have 0 as their righthand digit.
Multiples of 3 are such that the sum of the digits is also a multiple of 3.

The multiplication table on any other planet can be inspected similarly.

10

Multiplication
of Large Numbers

The traditional approach to 'getting the answer' to a multiplication computation looked like this when complete:

```
        729
   x     64
      2916
      4374
     46656
```

The only variation seemed to be that a zero (0) could be placed in the units place of the second sub - product, to read 43,740. Often the 0 was written before the other figures in the same line. Otherwise there was a shift to the left, carried out in a rote manner, so that the 4 turned up as it should in the 'tens position'.

More recently in schools students have been encouraged to list the subproduct standards:

```
        729
   x     64
         36
         80
       2800
        540
       1200
      42000
      46656
```

Both algorithms depend for their understanding on the distributive principle though perhaps the second version indicates more clearly the student's awareness of this. For mastery, the distributive principle and its application is necessary. The principle can be understood directly,

1. from one of the patterns in the multiplication table:
   Consider any two numbers in a row. Is the sum of these numbers also in the same row? Somewhere, even if not written!
   e.g., row 5: $20 + 30 = 50$
   The corresponding product forms would be,
   $(5 \times 4) + (5 \times 6) = 5 \times 10$
   $$= 5 \times (4 + 6)$$
   An endless number of examples of the distribution principle can, similarly, be extracted from the Multiplication Table.

2. By using rods and setting out four trains, each of length 38:

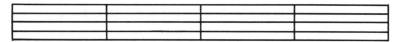

   This is equivalent to four thirties and four eights. We write $4 \times 38 = 4 \times (30 + 8) = (4 \times 30) + (4 \times 8)$.

3. Using counters we have four piles, each of 38. They can be subsetted in various ways. In particular, if groups of 30 are formed then, as before,
   $$4 \times 38 = 4 \times (30 + 8) = (4 \times 30) + (4 \times 8)$$

4. The spoken language has implications of the distributive principle. "Four times – thirty eight" sounds different from "Four times thirty – eight;" "Four times — thirty plus eight" different from "Four times thirty — plus eight." The position of the pauses implies different interpretations.

5. If students tend to write $4 \times (30 + 8) = (4 \times 30) + 8$ it is likely that the stress is upon the proximity of the written 4 and 30, not associating the multiplier 4 also with the 8, further away. This seemed in the past to be a common error when more emphasis was laid on the written form rather than on the substance of meaning. If the substitution of number names process is by this time second nature to the children, such mistakes are very unlikely.
   For if $4 \times (30 + 8) = (4 \times 30) + 8$

   then surely $4 \times (8 + 30) = (4 \times 8) + 30$

   But $30 + 8 = 8 + 30$

   so apparently $4 \times (30 + 8) = 4 \times (8 + 30)$ and $(4 \times 30) + 8$ would equal the $(4 \times 8) + 30$.
   This, however, is simply not true! So the original equivalence must be incorrect, namely $4 \times (30 + 8)$ *cannot* equal $(4 \times 30) + 8$!

Subproducts

There are many alternatives to the distribution in a product name.

$4 \times 38 = 4 \times (10 + 10 + 10 + 8)$
$\qquad\quad = 4 \times (20 + 10 + 8)$
$\qquad\quad = 4 \times (40 - 2) \ldots$

Students can be challenged to invent scores of them. Study of patterns in these is part of the mastery. For the first example above there are 4 subproducts, as seen by the four sub - rectangles in the diagrams of rods in no. 2 above. For the second there are 3 subproducts, for the third 2 subproducts.

What happens if we have:

(2 term non - standard) x (2 term non - standard)?

Consider, for example, 36 x 24.

$36 \times (20 + 4) = (36 \times 20) + (36 \times 4)$

However, $36 \times 20 = 20 \times 36 = 20 \times (30 + 6) = (20 \times 30) + (20 \times 6)$
$\qquad\quad 36 \times 4 = 4 \times 36 = 4 \times (30 + 6) = (4 \times 30) + (4 \times 6)$
from what has just been developed.

Therefore, 36 x 24 = (20 x 30) + (4 x 30) + (20 x 6) + (4 x 6) though, as usual, there are many alternatives to be sought and practised. The order of the 36 and the 24 in the product can be changed. So can the order of the *sum* of the subproducts or any or all of each subproduct term.

The standard is finally obtained by adding the subproducts. Thus, briefly put,

(2 term) x (2 term) gives 4 subproducts
similarly (2 term) x (3 term) gives 6 subproducts
(2 term) x (4 term) gives 8 subproducts
and (n term) x (m term) gives n x m subproducts

Mastery of all this provides another essential for multiplication computations and the essence of this is algebraic, applying to *all* examples.

Multiplication of large numbers

For those who wish to aim at speed in the multiplication of large numbers or at least see what can be done by emphasis on pattern usage, the following may be of interest.

Consider, first, the product of two 2 - figure numerals, 83 and 46 for example. There will be four subproducts if we substitute 80 + 3 for 83, 40 + 6 for 46. Traditionally, we have

$$
\begin{array}{r}
83 \\
\times\,46 \\
\hline
498 \\
3320 \\
\hline
3818
\end{array}
$$

Suppose we tackle the computation this way:

$$
\begin{array}{r}
83 \\
\times\,46 \\
\hline
_1\,8
\end{array}
$$

a) What subproduct fixes the unit digit in the standard name? (Answer: 6 x 3. "Put down 8, carry 1")

b) Which subproduct fixes the tens digit?

83
46
₆ 18

6 x 8, because that is really 6 x 8 - ty (or 6 x 8 *tens*). But 4 (- ty) x 3 also contributes to the tens digit. Each of these is standardized and added. That's 48 + 12 = 60. The 1 carried gives 61 (really 610). "Put down 1, carry 6."

c) Which subproduct fixes the hundreds digit? 4 x 8 because it represents 40 x 80, or 32 *hundred* .

The eye can follow these paths in effecting all this:

86
x 43

The first subproduct is 3 x 6.

86
x 43

The next two, 3 x 8 and 4 x 6 are added together with any carried number of tens (1 in this case).

86
x 43

Lastly, the subproduct 4 x 8 gives the hundreds digits, not forgetting any number carried from the previous step.

Notice the pattern represented by the short lines between the dots indicating the separate digits.

For 3 digit number names the pattern is extended:

586
x 243

identifies the unit digit.

together with any number carried gives the tens digit.

together with any number carried gives the hundreds digit.

together with any number carried gives the thousands digit.

together with any number carried gives the ten thousands digit.

The dot pattern can be examined for itself.

Notes

1   The total of the subproducts is 9, as expected, a square number (3 x 3 digits).

2

3   The distribution of the number of subproducts in each set of drawings is symmetrical (up and down identical). (Interesting that 1 + 2 + 3 + 2 + 1 is 3 squared. Will 1 + 2 + 3 + 4 + 3 + 2 + 1 be 4 squared? etc.)

2

1   The middle step (3 subproducts summed) will be the hardest step to compute, because more numbers have to be added together at this stage.

9

After the middle, the steps follow the same pattern as previously, but the steps are shifted to the left.

The whole drawing can be rotated through 2 right angles around the centre point of the to reappear as before. The drawing is, therefore, an example of 'point symmetry'.

Larger and Larger Products

It is easy to predict the patterns for the product of two 4 - digit numerals, or 5 digits or larger. The patterns have only to be extended. Given an example, therefore, the pattern of eye movements is superimposed on the written numerals. Practice will be needed in the actual substitutions before a learner is so sure of his accuracy that he will risk this new skill to amaze those not in the know, but given *that* , many intermediate grade children will be able to write down the standard answer to products of five or more digit numerals *without any recorded working!*

We give one example in detail, for a 4 x 4 product, realizing that in print – to be read – it all looks much more difficult than it really is. If a reader likes to tackle an example for himself he is recommended at first to get the pattern correct, not worrying at all about accuracy. Once the laying on of the steps of the pattern is easy for a learner, the number name substitutions can be stressed.

5384		7 fours is <u>28.</u> Put down 8, carry 2.
x 2697		
8		
		7 eights is 56, plus 2 is 58
		9 fours is 36, and 58 is <u>94.</u>
		Put down 4, carry 9.
48		7 threes is 21, and 9 is 30.
		9 eights is 72, and 30 is 102.
		6 fours is 24 and 102 is <u>126.</u>
		Put down 6, carry 12.
648		7 fives is 35, and 12 is 47
		9 threes is 27, and 47 is 74.
		6 eights is 48, and 74 is 122.
		2 fours is 8, and 122 is <u>130.</u>
		Put down 0 and carry 13.
0648		9 fives is 45, and 13 is 58.
		6 threes is 18, and 58 is 76.
		2 eights is 16, and 76 is <u>92.</u>
		Put down 2, carry 9.
20648		6 fives is 30, and 9 is 39.
		2 threes is 6, and 39 is <u>45.</u>
		Put down 5, carry 4.
520648		2 fives is 10, and 4 is <u>14.</u>
		Put down 14.

14520648 $\begin{matrix} \bullet & \bullet & \bullet & \bullet \\ \bullet & \bullet & \bullet & \bullet \end{matrix}$

It is not intended that any working be written, the internal verbalizations needing less energy. Words can be saved, for instead of saying to oneself, for example, "7 eights is 56, plus 2 is 58" one can get into the habit of saying "7 eights is . . . 58". No one else hears!

NOTE: The above process apparently deals with two numerals with the same number of digits. One may well ask, "What about 5384 x 697, 5384 x 97, or even 5384 x 7?". The answer is that the same process *can* be used because 697, 97 and 7 can be viewed as four - digit numerals, 0697, 0097 and 0007. Obviously, however, no one *would* use this method for 5384 x 7!

11

Division
Another Creative Activity

Multiplication was, in chapter 8, initiated by adding together sets of the same size (equivalent sets), represented, perhaps, by counters or fingers. Inversely, we may begin with, say, 64 counters and study the activity of repeatedly taking away eleven of the counters.

> "I begin with 64 counters in the pile. When I take away 11 there are 64 - 11, or 53 counters left in the pile. If I remove another 11 I have taken away 2 elevens; the remainder is 42. If I remove another 11 I have taken away 3 elevens; the remainder is 31. If I remove another 11 I have taken away 4 elevens; the remainder is 20. If I remove another 11 I have taken away 5 elevens; the remainder is 9. I cannot take away another 11 counters; there aren't enough left".

Some such words as these can accompany the activity of a child studying the situation. Instead of a long *written* record, however, a primitive writing can be used, only the successive numbers in each case being used:

$$64 \quad 11 \quad 2 \quad 42$$
$$64 \quad 11 \quad 3 \quad 31$$
$$64 \quad 11 \quad 4 \quad 20$$
$$64 \quad 11 \quad 5 \quad 9$$

Alternatively, the traditional convention can be introduced as shown in the following section.

Colored Rods

If we use colored rods we can request students to make, with white as the measure, a train equal to sixty - four whites.

We then ask, "Can you show sixty - four minus eleven? Minus another eleven? Can you subtract *three* elevens? What do you observe?"

The student replies that he certainly *can* place three elevens at the side of the sixty - four length and that there is still a length of thirty - one whites unfilled.

The teacher says this can be recorded simply as:

$64 \div 11 = 3$ and 31 remain
or $64 \div 11 = 3$ and 31 remainder
or $64 \div 11 = 3$ and rem.

Immediately following either of these activities the students can be asked: "Could you subtract *more* than three elevens?"

Either as a result of previous activity, or the implication of finding other possibilities *now,* someone reports that *he* can subtract *four* elevens. Pinned to a precise description this pupil says, or writes:

$64 \div 11 = 4$ and 20 rem.

Someone else says:
$64 \div 11 = 5$ and 9 rem.

The teacher may also question the completion of statements like:

$64 \div 11 = 2$ and ? rem.
and
$64 \div 11 = 1$ and ? rem.
and especially
$64 \div 11 = 0$ and ? rem.

Experience suggests that most children from 2nd and 3rd grade onwards will have little difficulty in establishing a set of statements:

$64 \div 11 = 0$ and 64 rem. $64 \div 11 = 0$ and 64 rem.

$64 \div 11 = 1$ and $(64 - 11)$ rem.   or with $64 \div 11 = 1$ and 53 rem.

$64 \div 11 = 2$ and $(64 - 11 - 11)$ rem.   standard $64 \div 11 = 2$ and 42 rem.

$64 \div 11 = 3$ and $(64 - 3 \times 11)$ rem.   names $64 \div 11 = 3$ and 31 rem.

$64 \div 11 = 4$ and $(64 - 4 \times 11)$ rem. $64 \div 11 = 4$ and 20 rem.

$64 \div 11 = 5$ and $(64 - 5 \times 11)$ rem. $64 \div 11 = 5$ and 9 rem.

Before all these statements are accepted by every member of the class the teacher can suggest that the students look for patterns. There is, for example, the 0, 1, 2, 3, 4, 5 sequence of numbers, and that of 64, 53, 42, 31, 20, 9. Do these diminish by eleven each time and is this surprising? Can every child see that when we remove eleven pebbles from the table the remaining pebbles *must* diminish by eleven? Or that as every train of rods, eleven in length, is placed in line with the other elevens, the remaining space up to the end of the sixty - four train *must* lessen by eleven? Is it reasonable for the teacher to expect every child to see this and see it as much simpler than the appropriate explanation in words?

Why did we choose 64 and 11 as the numbers in this division activity? Does it matter what numbers *are* chosen, in order to understand the process?

For first grade introduction it would be wise to use, as the first criterion of choice, numbers familiar to the student. It could be perhaps $7 \div 2$ or $7 \div 3$ but if *they* are manageable then so is,

$$(7 + 7) \div 3$$
$$\text{and } (7 + 7 + 7 + 7) \div 3$$

so long as one did not necessarily want the *standard name* for $7 + 7 + 7 + 7$. By using pebbles or rods, every student could, provided he can count to seven, tackle $(7 + 7 + 7 + 7) \div 3$. Maybe his report would be that:

$$(7 + 7 + 7 + 7) \div 3 = 2 \text{ and } (1 + 7 + 7 + 7) \text{ remainder}$$

or perhaps

$$(7 + 7 + 7 + 7) \div 3 = 2 \text{ and } 1 \text{ rem.}, + 2 \text{ and } 1 \text{ rem.}, + 2 \text{ and}$$
$$1 \text{ rem.}, + 2 \text{ and } 1 \text{ rem.} = 8 \text{ and } 4 \text{ remainder}$$

Already many possibilities are attainable by the students, provided they are not expected to write standard names. If they happen to know some standard names, all well and good – then there will be more ways of reporting or recording what is seen as a result of the activity.

Generally then, it is recommended the early activities be with fairly small numbers of pebbles, but it is wise to remember that if the numbers are too small, the activity can be insipid and communication may be more difficult. Better to deal with the really special small number cases later, after the basic rules of play are established.

$66 \div 11$ or $64 \div 11$?

Is it better to begin with examples such that one number is a factor of the other, one divides the other – 'evenly' as the jargon goes? Or is it preferable to have a remainder?

We choose the latter, as we can then say there is *always* a remainder, even if sometimes it is zero. Not that we suggest that there is anything inappropriate in examples such as $12 \div 2$, $12 \div 3$, $16 \div 8$. The difficulty arises when, having had a surfeit of these, the student meets $12 \div 5$ as something new and queer. Rather, we may need to sacrifice more easily obtained 'answers' in the early stages when we are seeking more generalized ways of looking at the situation, which are valid regardless of the numbers chosen.

Relationship between x and ÷

The fact that we write:

$$64 \div 11 = 3 \text{ and } 31 \text{ remainder,}$$

(choosing the example already used) does not mean that we necessarily think of the operation of division when we observe a person performing the activity with the pebbles or the rods. For, regardless of the thoughts of the person in the activity, an onlooker might think along other lines:

> I notice that 11 pebbles were picked up three times and that there were then 31 pebbles remaining on the table. So I say that '3 times 11' plus 31 must equal 64. And I write:
> $$31 + (3 \times 11) = 64.$$

Thus the two statements

$$64 \div 11 = 3 \text{ and } 31 \text{ remainder}$$
$$\text{and } 31 + (3 \times 11) = 64$$

are interdependent. One is a *transformation* of the other and we want students to see this interdependency.

It is similar with the rods. If the arrangement of rods is like this:

10	10	10	10	10	10	4
11	11	11	31			

either of the two above statements is legitimate. Each is a record of one of the viewpoints possible in looking at and considering the pattern of rods.

This relationship between a statement concerning division and one concerning multiplication is one which will be constantly used and made conscious when we are working on division and multiplication. Sometimes this can be utilized as a checking device, but fundamentally it is inseparable from an understanding of multiplication and division. Analogous understandings will follow in operations upon integers, rational and irrational numbers.

Back to 64 ÷ 11

There are appearing now, from the simple beginning of division, so many possible ways of proceeding that it is impossible to say what precise order should or should not be followed. The criteria which decide this suggest that all that has been done is basically appropriate to all grades (with certain conditions, the most important being the assurance of the teacher that standard names are *not* particularly important until the operations are mastered).

However, another path to follow can arise from further consideration of:

$$64 \div 11 = 0 \text{ and } 64 \text{ rem.}$$
$$64 \div 11 = 1 \text{ and } 53 \text{ rem.}$$
$$64 \div 11 = 2 \text{ and } 42 \text{ rem.}$$
$$64 \div 11 = 3 \text{ and } 31 \text{ rem.}$$
$$64 \div 11 = 4 \text{ and } 20 \text{ rem.}$$
$$64 \div 11 = 5 \text{ and } 9 \text{ rem.}$$

For we can ask, having written on the board:

$$64 \div 11 = 6 \text{ and } ? \text{ rem.}$$

"What remainder would we have this time?"

If this is thought of in terms of the pebbles, an answer of 'zero' might be forthcoming, for having taken away five of the subsets of eleven pebbles, it is impossible to take away another subset of eleven. However, other answers might be given. With the rods, on the other hand, another 'eleven train' *can* be placed at the end of those already there, and recording what one has done would lead to something like: "I have subtracted six of the elevens, and have to *come back* 2 to complete the 64 length."

10	10	10	10	10	10	4
11	11	11	11	11	11	

This can be written:

$$64 \div 11 = 6 \text{ and 'come back 2'}$$

or some students sometimes say,

$$64 \div 11 = 6 \text{ and '2 overhang'}.$$

If the teacher wishes, he can say that we usually call this 'negative two', write it $^-2$, and everyone writes:

$$64 \div 11 = 6 \text{ and } ^-2 \text{ remainder}$$

(although the word 'remainder' would hardly be used if it had not been used in the first place).

Usually the flood gates now open and the students continue:

$$64 \div 11 = 7 \text{ and } ^-13 \text{ rem.}$$
$$64 \div 11 = 8 \text{ and } ^-24 \text{ rem. and so on.}$$

Of course, all students will not necessarily write $^-2$, $^-13$, $^-24$ ... successively. But even if they don't, they can still write the corresponding symbols:

$$9-11, 9-11-11, 9-11-11-11 \ldots$$

or perhaps $9-11, 9-2 \times 11, 9-3 \times 11 \ldots$

or perhaps "2 below zero, 11 below 2 below zero, 11 below 11 below 2 below zero ... etc."

We are not seeking, at this time, standard names or sophisticated operations with negative numbers. The emphasis is the seeking of patterns and reasonable predictions to fit in with them.

Large Numbers

If, using the pebble activity, we consider,

$$64 \div 11$$

and our report so far is that,

$$64 \div 11 = 4 \text{ and 20 rem.}$$

what would happen if we began the activity again, still removed four subsets each of eleven pebbles, but had on the table initially not sixty - four pebbles but one hundred more? Would we not be able to perform identical activities but have a greater number of pebbles remaining on the table? And would not the remainder be increased by the one hundred? Thus:

$$164 \div 11 = 4 \text{ and 120 rem., or}$$
$$164 \div 11 = 4 \text{ and } (100 + 20) \text{ rem.}$$

If, instead of one hundred extra we had, shall we say, eighty - seven extra wouldn't we say that,

$$(64 + 87) \div 11 = 4 \text{ and } (20 + 87) \text{ rem.?}$$

Do we not want children, therefore, to see that regardless of the increasing number of pebbles, true division statements can still be made? We can even get the students to understand that another million pebbles added to the sixty - four would not deprive us of the insight that:

$$(64 + 1,000,000) \div 11 = 4 \text{ and } (20 + 1,000,000) \text{ rem.}$$

Even if the children do not know how to write a million, or they write it incorrectly, they can *see* the generalization and talk about it.

In this sense the size of the numbers involved in the division operation does not hinder progress. On the contrary, they help us understand the 'algebra of the situation', providing we are not seeking standard names as answers. The latter aim, to be sure, is one frequently to be pursued but is appropriate only as sophistication increases with knowledge of a greater vocabulary – what has commonly been called the 'number facts' or 'the tables'.

In detail, we suggest that:
$(64 + 1,000,000) \div 11 = 4$ and $(20 + 1,000,000)$ rem., or
$(64 + 98,274) \div 11 = 4$ and $(20 + 98,274)$ rem.

are *easy* to understand and arise from considerations of activities with pebbles or rods. We may even expect some extra progress such as
$(64 + 1,000,000) \div 11 = 5$ and $(9 + 1,000,000)$ rem., or even
$(64 + 1,000,000) \div 11 = 6$ and $(^- 2 + 1,000,000)$ rem.

This is not, however, to imply that children can necessarily write $(64 + 1$ million) in the conventional numeral form, 1,000,064, let alone receive as a problem a collection of symbols like
$1,000,064 \div 11 = \square$
with the expectation by teacher that it will be understood and that from this form of open sentence alone the standard answer will be obtained!

$1,000,064 \div 11 = 7$ and 999,987 rem.
$1,000,064 \div 11 = 8$ and 999,976 rem.
$1,000,064 \div 11 = 90,914$ and 10 rem.

One day we certainly want this, but not yet. The present work is to lay the foundations for the rapid, efficient computational exercise as one later by - product of our mathematics. If the fundamental understandings are mastered the efficient computational skills can be built on them. Otherwise, most of what we do will be superficial. Anything remembered but not understood will have only restricted application and value.

Long Division

Division has been seen above as a matter simply of repeated subtraction, the same number being subtracted again and again. So, whether a student from his early school days gradually accumulates an experience of division activities first from physical actions and then by symbolizing these actions, or whether the student is older at perhaps fifth or a later grade, the minimum we ask is that he realizes that division can be interpreted as stated. Thus, asked to divide 673 by 26 we are minimally content if a learner tackles the computation implied thus:

$$\begin{array}{r} 673 \\ - 26 \\ \hline 647 \\ - 26 \\ \hline 621 \\ - 26 \\ \hline 595 \end{array}$$

At this point the result can be stated as
$$673 \div 26 = 3 \text{ and } 595 \text{ rem.}$$

Even though this may be the first time the student has done this it will probably not be long before he begins to seek short cuts. Discussion of other students' activities, suggestions from the teacher, lead to a realization that others may write this:

$$
\begin{array}{r}
673 \\
- \ 52 \\
\hline
621 \\
- \ 52 \\
\hline
569 \quad \ldots \text{etc.}
\end{array}
$$

Fifty-two is repeatedly subtracted because this student *happened* to know that $2 \times 26 = 52$ and it came to his mind for use in this particular context. Other examples gleaned from the class will probably reveal other developments:

$$
\begin{array}{r}
673 \\
- \ 130 \\
\hline
543 \\
- \ 130 \\
\hline
413 \\
- \ 130 \\
\hline
283
\end{array}
\qquad
\begin{array}{r}
673 \\
- \ 52 \\
\hline
621 \\
- \ 130 \\
\hline
491 \\
- \ 260 \\
\hline
231 \\
- \ 130 \\
\hline
101
\end{array}
\qquad
\begin{array}{r}
673 \\
- \ 260 \\
\hline
413 \\
- \ 260 \\
\hline
153 \\
- \ 130 \\
\hline
23
\end{array}
$$

$673 \div 26$
= 15 and 283 rem.

$673 \div 26$
$= (2 + 5 + 10 + 5)$ and
101 rem.
$= 22$ and 101 rem.

$673 \div 26$
$= (10 + 10 + 5)$ and
23 rem.
$= 25$ and 23 rem.

The last written algorithm is that which is frequently used in modern textbooks though slightly amended:

$$
\begin{array}{r|r}
26) \ 673 & \\
260 & 10 \\
\hline
413 & \\
260 & 10 \\
\hline
153 & \\
130 & 5 \\
\hline
23 & 25
\end{array}
$$

$673 \div 26 = 25$ and 23 rem.

What might be questioned about this, however, is whether the 'partial quotient' using 10's is *necessarily* the easiest for everyone, and whether it is necessary for children in their early division experience to set out the algorithm in this conventional way, writing vertically down the sheet of paper?

To both implied queries we answer 'no'. Many children, given the freedom to use whatever subtractive steps they wish to use, do *not* choose multiples of ten every time. Sometimes even multiples of nine are preferred:

$$
\begin{array}{r}
673 \\
- 234 \\
\hline
439 \\
- 234 \\
\hline
205 \quad \ldots \text{etc.}
\end{array}
$$

We are led to conclude that in this case, this particular student happened to bring to mind easily and rapidly the fact that subtracting 234 was an operation equivalent to that of subtracting nine 26's one after the other.

It is interesting to note that there is no dramatic quick way of processing the standard answer for division as there was for multiplication. The standard is that result in which the remainder is as small as possible without being negative. Always a 'trial and error' process is necessary, but the power of our development here provides understanding of what is basic which with creative flexibility provides avenues to at least non - standard equivalents regardless of what numbers are in the examples met. Standard answers will come as important by - products when needed experience in the process is gained.

The Old Approach

Compare what we have outlined with what used to be done:

$$26)\overline{673}$$

We would have had the student say something like this, with little or no experience beforehand with physical objects or sets:

"Twenty-six into six, won't go . . .

> (Despite the fact that the student invariably reads 673 as 'six *hundred* seventy-three' he was now required to say that the first digit was *six* ! Why shouldn't it again be named 'six hundred'?)

Put a zero (or dot) over the 6.

Twenty - six into sixty - seven . . . try three times . .

$$\begin{array}{r} 26 \\ \times\ 3 \\ \hline 78 \end{array}$$

. . . seventy - eight . . . too large . . . try two times . .

$$\begin{array}{r} 26 \\ \times\ 2 \\ \hline 52 \end{array}$$

. . . that's better . . . write the 52 under the 67 and subtract,

$$\begin{array}{r} 0 \\ 26)\overline{\,673} \\ 52 \\ \hline 15 \end{array}$$

. . . put the 2 over the 7. Bring down the 3,

$$\begin{array}{r} 02 \\ 26)\overline{673} \\ 52 \\ \hline 153 \end{array}$$

. . . twenty - six into one hundred fifty - three . . . try five times . . .

$$\begin{array}{r} 26 \\ \times\ 5 \\ \hline 130 \end{array}$$

$$
\begin{array}{r}
025 \\
\hline
26\,)\,673 \\
52 \\
\hline
153 \\
130 \\
\hline
23
\end{array}
$$

<u>Answer:</u> 25 and 23 remainder.

Sure, we can get the 'right answer' by this method – if we remember exactly how the set - up on paper goes. But at what a cost! Creativity and personal choice are absent. Of course, there may indeed be *some* understanding of the underlying concepts although the evidence of the present generation of adults raised on this method leads one to doubt this. It was the *form* of the written algorithm which was all important in days gone past. Now we seek some substance as well, from which a variety of acceptable written forms can emanate.

12

Fractions
are not Parts of Wholes

Historical Survey

The topic of fractions is well established in the later years of the elementary schools and continues throughout the secondary grades under headings like: 'algebraic fractions' and 'trigonometrical fractions'. A great deal of time and energy seems to be spent on the topic, especially in the sixth and seventh grades at which time all the fraction operations are dealt with. Despite this, however, there is evidence that the understanding of the basic concepts of fractions with the skills of manipulation are not as successful or rewarding as teachers might wish, considering the time, energy and devotion involved. Secondary school pupils, university students and teachers themselves see fractions as a major obstacle in their mathematical experience and although they may well retain a few isolated rules of "how to do multiplication or addition", the reasons for such rules are rarely understood.

It may be valuable to consider at first the assumptions on which fractions have been taught in schools in years past and then see whether today any alternatives are available.

Some fraction words and phrases abound in common speech and most children know and use these relevantly during their earliest school years. 'Half an hour', 'quarter of a kilometre', 'a half day' – these are easily accepted and so too, are simple examples of operations: 'two quarters for a half dollar' or 'two thirds of the hockey match are over, one third to go'.

Historically, it seems plausible that it was the use of such everyday phrases by children which caused teachers of the 19th century to begin a more formal look at fractions in school with examples involving halves and quarters:

$$\frac{1}{2} + \frac{1}{2} = 1$$
$$\frac{1}{2} + \frac{1}{4} = \frac{3}{4}$$

working up from these to examples considered harder like:

$$\frac{1}{8} + \frac{3}{8} = \frac{4}{8} = \frac{1}{2}$$

Problems involving 'tenths' perhaps 'twelfths', 'sixteenths' and 'thirty‑seconds' came next, on the grounds that these would be useful for those students who contemplated going into jobs involving measurements – wood or metal work, commerce or minor engineering.

The usefulness for a future job was an important criterion in those days for determining what was presented to children. It would be absurd, the argument went, to use fractions such as $\frac{4}{13}$, $\frac{14}{27}$, $\frac{1}{9}$, for they would never be used in practice. A similar argument that some aspects of language teaching or story telling were acceptable for children even though they might not come in useful in the future, was not made. Children read and said nursery rhymes not because they would be useful but because they were enjoyable and firmly established by tradition in the homes. Fraction talk was not!

As fractions were introduced more and more into classrooms, teachers did their best to communicate what they were all about to their young pupils. Few teachers by 1900 would have had formal or mathematical definitions. They were affected by the practical application pressures and adapted these for school exercise in measurement and in the use of 'pies'.

'Pies', of course, were circles drawn on paper showing slices of fourths, eighths, and so on. This provided a visual referent for the students to help them learn, and the use of fraction pies became firmly established as a valid method for aiding communication.

Also firmly established, however, was the impression that a fraction was 'part of a whole'. To this day that phrase is the most common working definition heard in the classroom.

Now if the phrase 'a part of a whole' is merely either jargon which serves the purpose on occasions of signalling a topic for deliberation among people who already understand or if it represents some application to measurement or money, based again on understanding, then the phrase plays a useful role. But commonly the phrase is used at the beginning of the understanding, so we must ask ourselves what advantages and possibly disadvantages there are if a student either takes the phrase literally or studies too closely the analogy of the pie.

Advantages would seem to include the fact that pies are easily illustrated, they can be sectioned by drawing lines through the centre, and they cost nothing as learning aids.

Disadvantages need more study. One can show with a drawing a representation of 'three fourths' of a pie, but one cannot show 'four thirds'! One can, of course, show 'three thirds' of one pie and 'one third of another' so long as one tacitly accepts that the pies are equivalent (congruent, in fact). We cannot say this difficulty alone caused the name 'improper fraction' to be used, but most certainly vast numbers of students feel about fractions greater than 1 differently from what they feel about 'proper fractions'. 'Improper fractions' *have* to be written as mixed fractions; it is unacceptable to leave them as they are – these are common reactions in school.

Next, the shape of pies seldom varied so we can believe that a very rigid conditioning resulted from the constant use of circular drawings. The circles could be divided into halves, quarters, eighths, twelfths – as on a clock face – but not so easily into fifths, sevenths, and elevenths.

The symmetry of the circle is at the same an advantage and a disadvantage; advantage because with lines through the centre the circle is easily sectioned, disadvantage because if only circular 'pies' are used there may well be too rigid an association between the idea of fraction and the sector shape of each 'piece of pie'.

Of course, much more can be done with 'pies' as a learning aid, but it seldom is. Different shapes can be used: squares, triangles, other polygons and, rather than illustrate those fractions which best lend themselves to a symmetrical sectioning of the shape, challenges to invent others can be made. A square, for example, lends itself visually to an application of fourths.

 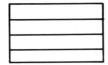

What can we do for 'fifths' and other fractions? In how many ways can we show 'tenths' of a square? Are these all valid?

 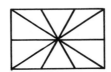 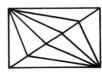

But as one attempts to use more and more fractions, or more and more shapes, difficulties arise. Some of the simplest drawings of pie sections are acceptable because the eye accepts the equivalence of the pieces. Others are more obscure and may depend on extra investigation – not that this is to be deprecated.

Lastly, a disadvantage of the usual treatment is that too seldom the student is asked to do anything but look – look at what may be a complicated drawing done by someone else, capable of ambiguous interpretations. If the learners are encouraged to cut out shapes for themselves, all well and good. They would then be involved in actions from which they could draw their own conclusions. This is seldom done, however, probably because the drawing of the sections of shapes is rarely satisfying or satisfactory, even if we imagine greater tolerance for pencil work than is usually present in classrooms.

An alternative approach *is* available – though it does not mean the sectioning of shapes becomes completely irrelevant. That becomes, not a method of learning about fractions, but one of the applications of the basic learning when it has taken place.

The alternative is to accept fractions as understood and used by mathematicians and to embed this within situations so that students involved are most likely to draw conclusions according to such understanding. Rather than using 'parts of wholes' we suggest fractions are symbols which are used for rational numbers and they, the rationals, are seen as families of pairs of whole numbers.

Rational Numbers

We have already met various families of pairs of whole numbers. The family, some of whose members are: (2,1), (3,2), (4,3) and so on, was called the relation 'one more than'; that whose members include such pairs as (2,1), (4,2), (6,3), (8,4), can be called 'double' or the 'doubling relation'.

A simple extension of this process will give (4,7), (8,14), (12,21), (16,28), as members of another family with the understanding that as many other members can be generated at will according to the basic pattern: that the numbers in every pair have a highest common factor and that the complement of these are 4 and 7. In other words, every member of the family can be represented in the form (4 x n, 7 x n) where *n* is any whole number (although not zero).

Another insight into the condition of membership of this family would be to say that any two ordered pairs of whole numbers belong if the product of the first number of the first pair and the second number of the second pair is equal to the product of the second number of the first pair and first number of the second pair. In short: if (a,b) and (c,d) are to be members of the family given, it must be true that a x d = b x c where a, b, c, d are whole numbers. This *is* true for the members already quoted:

> (4,7), (8,14), (12,21), (16,28)
> (4,7) and (8,14) are members of this family because 4 x 14 = 7 x 8
> (4,7) and (12,21) are members of this family because 4 x 21 = 7 x 12
> and so on.

This family is called 'four sevenths' or the 'rational number four sevenths'. There is an infinite number of members of the family, for we can have (8 x 4, 8 x 7), (9 x 4, 9 x 7), ... (27 x 4, 27 x 7), ... (35 x 27 x 4, 35 x 27 x 7) as members. Clearly there is no end to the generation of more pairs by this process just as there is no end to the numbers available as multiples of both the 4 and of the 7.

If the written mark $\frac{4}{7}$ is used to represent the rational number, it is the mark, itself, which is called the fraction. If, alternatively, (4,7) is used, we can call this the 'ordered pair' form of representation. It is not easy to use these terms consistently for in our culture ambiguity has already been established. $\frac{4}{7}$ does not always represent a rational number but it *is* the commonest representation used in elementary and secondary schools.

Initially, then, it is recommended that work with the fractions used to represent rational numbers, apart perhaps from some early casual application of vocabulary as already cited, should be built on activities based on the appreciation of families of ordered pairs of whole numbers, the nature of such families, and the process by which members of any fraction family can be generated easily at will. Given this understanding mastered, the operations called addition, subtraction, multiplication and division become simple, both in understanding and in skills.

The Equivalence of Fractions

Using Colored Rods

We have had much experience in measuring one rod or a train of rods using any other rod (or train of rods) as the unit. Sometimes we found that the measuring length was a factor of the measured length. In those cases we obtained a whole number. Thus, an orange rod measured with the length of a red rod as unit gave a number which we call 'five' and write 5. If the yellow length is the unit then 2 is obtained from the orange length.

$$\frac{2}{9} \times \frac{9}{7} = \frac{2}{7}$$

$$\frac{2}{11} \times \frac{11}{8} = \frac{2}{8}$$

$$\frac{18}{65} \times \frac{65}{112} \times \frac{112}{42} = \frac{18}{42}$$

$$\frac{24}{7} \times \frac{7}{8} \times \frac{8}{64} = \frac{24}{64}$$

$$\frac{9}{100} \times \frac{100}{99} \times \frac{99}{4} \times \frac{4}{3} = \frac{9}{3}$$

$$\frac{4}{3} \times \frac{3}{6} \times \frac{6}{99} = \frac{4}{99}$$

$$\frac{17}{1,000} \times \frac{1,000}{2} \times \frac{2}{108} =$$

$$\frac{20}{4} \times \frac{4}{5} \times \frac{5}{2} = \frac{20}{2}$$

$$\frac{14}{102} \times \frac{102}{9} \times \frac{9}{1} = \frac{14}{1}$$

$$\frac{7}{2} \times \frac{2}{9} = \frac{7}{9}$$

Multiplication of fractions

We can now extend this process. If we, say, measure the 'orange + dark - green train' with red lengths we get 8; measuring the 'orange + orange + brown train' also with red rods we get 14. So if we measure the pair of trains,
$$(o + d, o + o + t) \qquad t \text{ for 'tan' or brown}$$
using the red as the unit, we get the pair of numbers (8,14), writing it thus, if we wish.

But *any* length can be used as a unit – there is certainly no reason always to choose the red. We might choose the white to measure the same two trains. If we do, we get the pair of numbers (16, 28).

We are driven to the conclusion, therefore, that the relationship between the lengths of the two trains being considered cannot only be written as (8,14), but also as (16,28).

This activity can be repeated with the pink rod as the unit. The pair (4,7) emerges. Thus, the three pairs (4,7), (8,14), (16,28) are all names for each other; they belong to the same 'family' and we can write:
$$(4,7) = (8,14) = (16,28)$$

An alternative way of writing (4,7) is as the fraction, $\frac{4}{7}$. This is the usual convention, but it need not represent anything other than what has been understood so far.

Other members of the family 'four sevenths' can be found by a number of means. The measuring unit can be changed as has been indicated. If this seems to limit progress we say the white represents 2 simply by assigning the name 'two'. Measurement could be 3, 4, or any number one wishes – or the student wishes – and new members obtained: (3 x 16, 3 x 28), (4 x 16, 4 x 28) and so on.

Alternatively we might line up two trains, the first: four blue rods, say, the second: seven blue rods. This will be an application of the same relationship (4,7), but these same trains remeasured with white units will give (4 x 9, 7 x 9). Four orange rods and seven orange rods will give (40,70) using white as units, or (20,35) with reds as units. Beginning with any rod or train and treating its length as the unit, we only have to consider four of such lengths and then seven of such lengths, measure each again with any suitable unit and we have another two whole numbers which, in order, provide us with another member of the family 'four - sevenths'.

Using Counters

As indicated previously, what can be extracted from actions with rods can also be extracted from sets of counters, though more ponderously and less efficiently. The mechanics of counting individual counters, certainly when more than ten or twenty are needed, is indeed a slower process than picking up appropriate rods. The commonly experienced need for recounting by everyone desiring to be correct, points to the lower efficiency.

Such disadvantages need to be experienced at first hand, however, so a few examples during the introduction of fractions may be appropriate. The procedure is similar to that with rods.

We can always begin by picking up two separate sets of counters. In each set they are counted, checked and the numbers recorded.

Assume now that it is possible, with each of the sets chosen, to arrange their members – the counters, in *pairs*. We count the number of pairs. Whether we now use the numbers of counters or the number of *pairs* of counters, we clearly shall have a representation of the relation between the numbers of members of the sets.

If we have eighteen counters in one set, twenty in the other, then (18,20) is a pair representing the sets. We also get (9,10) from considering pairs of counters. We say that the (18,20) relation is the same as the (9,10) relation, or that 'eighteen - twentieths' is another name for 'nine - tenths'.

Commencing with another two sets, we could get (18,24) at first and (9,12) from the pairing. But in this example we can also place the counters in sets of triples. This produces (6,8). Further, with the counters in subsets of sixes, we have (3,4). This time we have more equivalent pairs:

$$(18,24) = (9,12) = (6,8) = (3,4)$$

The richness of the crop of equivalent names like this will, of course, depend on the choice of the two particular sets. If the numbers of counters have many common factors lots of equivalent names can be obtained. But if we start with 17 counters and 20 counters we shall not be able to change the measuring subset (the unit) as flexibly as with 18 and 20, or 18 and 24. In class, therefore, a teacher might be well advised to prepare appropriate sets ahead of time – and the accuracy of his counting will be important!

An alternative activity, which *will* give rise to many equivalent names is to begin with any two whole numbers and apply each to equivalent sets of counters. For instance, we can choose a set of counters, *any* set – take five of these sets and also take seven of them. Suppose there are eight counters in each of the sets (all the sets are equivalent in number to that first chosen). Then (5,7) applied to the initial sets gives (5 x 8, 7 x 8) applied to the counters.

Now change the basic set, to one of 11 counters, say. (5,7) applied to sets equivalent to this gives (5 x 11, 7 x 11) referring to counters.

So (5,7) = (5 x 8, 7 x 8) = (5 x 11, 7 x 11), and one can generate many more equivalent names by, again and again, changing the basic set. In fact, actual counters need not be used. We may look at 5 small boxes, and 7 small boxes and *imagine* that each of the boxes always contains the same number of counters as every other box. The number of equivalent names is without limit.

Using Other Objects

Objects which possess attributes amenable to changes of unit provide perhaps more flexibility than do counters, although care is necessary that simplicity is not obscured by an overcrowding of impressions.

Consider textbooks, for example. Sets of them are common enough in every classroom and it is usual to accept any book of a set as equivalent to any other of the same set. Even though a particular student may claim one book as *his* and, therefore, as different from all the others, it is easy to come to an agreement that all the books of a set *are* equivalent to one another.

That means we agree to accept that any one book has exactly the same properties as every other of the set: same cover, same pages, same ink marks. Even if all these are not actually true they can be *imagined* to be so for the purpose of our coming activity. Any book can be accepted as a substitute for any other of the set.

Given two subsets of books from the same set, therefore, we can discuss what pair of numbers can represent the two subsets. While the first pair of numbers given by children may well be the number of books in each subset, there are many other possibilities and these can be increasingly sought either by teacher suggestion, or by following those of the students:

(5,9) might be the pair arising from considering the number of *books* in each subset;

(20,36), however, could come from looking at the number of *words*, (four) on the cover of each book of each subset, or

(60,108) from the *letters* on each cover and then from all the covers of the subsets, or

(238 x 5, 238 x 9) from the number of *pages* of the books in each subset.

In an exercise like this, one deliberately uses the creative imagination of the students. It matters little whether the numbers actually refer to the books in question, only that if some assumption is made then a certain conclusion follows.

e.g. (158924 x 5, 158924 x 9) is a pair in the family, if there are 158,924 letters in one book, and, of course, the same number of letters in every book. (Whether there are actually that number of letters is, for this purpose, irrelevant!)

An important emphasis in this activity is to refrain from any pressure that a *standard name* is needed for each whole number. There is not only no need to figure out the standard form for 238 x 5, used above, it gets in the way. What we wish to convey at this stage is that (N x 5, N x 9) is a member of the (5,9) family for *any* whole number value of N (except 0).

On the other hand, there are occasions when we can place the emphasis on getting the standard names or we can begin with the standard names and find out whether or not we have a pair belonging to this or that family.

Using the Multiplication Table

There are presentations which do not depend directly upon the use of material aids. Given any multiplication table, as for instance:

x	1	2	3	4	5	6	7	8
1	1	2	3	4	_5_	6	7	_8_
2	2	4	6	8	_10_	12	14	_16_
3	3	6	9	12	_15_	18	21	_24_
4	4	8	12	16	20	24	28	32
5	5	10	15	20	25	30	35	40
6	6	12	18	24	30	36	42	48

the leader can say, "Another name for the fraction 'five eighths' is 'ten sixteenths'."

As he does so he taps with a pointer the 5 and the 8 in the first row (not the heading) and then the 10 and the 16 in the second row. He then invites the onlookers to give another name for 'five eighths'. Following the pattern of movement he initiated, they tap in turn 15 and 24, then 20 and 32, and so on, working down the successive rows. When they exhaust the actual writings, the next numbers can be extrapolated or, if work has previously been done on extensions of any multiplication table in all directions, there will be little difficulty in producing, 'seven times five, seven times eight', 'eight times five, eight times eight' and so on, as more members of this fraction family.

Should the students also be familiar with the extensions into negative numbers, other pairs may arise like, ($^-$ 5, $^-$ 8), ($^-$ 10, $^-$ 16), (5 x $^-$ 3, 8 x $^-$ 3).

If in the table there is a row (and a column) of zeros, it may be thought that (0,0) is also a member of the (5,8) family. This can be investigated further in the traditional way, namely that (0,0) will be a member of the (5,7) family, the (4,7) family ... in fact, *every* family. We cannot have this, however – (0,0) cannot be a member of the (5,8) family *and* the (5,7) family, as the (5,8) and (5,7) families are *not* equivalent: they cannot share any pair as one of their members.

We are driven to the conclusion, therefore, that (0,0) must be an exceptional pair. It is so open to ambiguous interpretation, that we decide not to include it in our deliberations. It is *not defined.*

Summary

Whether we use rods, counters, the multiplication table, or begin with two equivalent products, we wish to emphasize strongly the understanding of rational numbers as families of ordered pairs of whole numbers (or of integers if negative numbers are included) is vital.

Each family has members. Each member is a pair of whole numbers, not *any* pair, but pairs such that certain connections – relationships – exist between the four numbers of every two pairs. If (a,b) and (c,d) are pairs of the same rational number family, then either a x d = c x b or, putting it another way, if there is a whole number n such that c = n x a (i.e., n is a factor of c and a is its complement), then d = n x b. If these statements are not true, (a,b), (c,d) do *not* belong to the same rational number family – though they may well be members of some other family.

There is an infinity of members of each rational number family and for each member of every family there is an infinity of ways of naming it and writing the name.

The order of the two whole numbers in each member, in each pair, is significant. (2,5) is *not* in the same family as is (10,4) because 2 x 4 ≠ 10 x 5; (2,5) and (4,10) are equivalent, they *are* members of the same family because 2 x 10 = 4 x 5.

If one member of a rational number family is such that it's first whole number is less than the second, this will be true for *every* member. If the two numbers of every one member are the same, then they will be equal to each other for every other member. This, in fact, will be the family (1,1).

A family can be identified by the name of any one of its members, although the abundance of an infinity of possibilities suggests the usefulness of a standard name. Thus: (2,5), 'two fifths', (4,10), 'four tenths', (36,90), 'thirty - six ninetieths'... and so on, can all be the names for the same rational number. The standard chosen, however, is (2,5) because this is the only member of this particular family for which the two whole numbers, 2 and 5, are prime to each other. In other words, they have no common factor (other than 1). In all other cases the two numbers of each pair *do* have common factors.

(2,5) is traditionally called the 'irreducible element' or 'the fraction in its lowest terms'. Note that (⁻2, ⁻5) can also be regarded as a member, the factor common to ⁻2 and ⁻5 being ⁻1. So, in a sense, (2,5) *can* be 'reduced' or its terms 'lowered'.

The family (2,5) is called the 'reciprocal' of the family (5,2). So the reciprocal of the reciprocal of a family is the family itself!

A meaning is given to whether one family is or is not 'greater' than another family, by identifying one member of each which share the same second whole number. This is always possible for we may choose as the second whole number the product of the two separate given second numbers. For example, for (4,6) and (3,10) we choose (4 x 10, 6 x 10) as an equivalent to (4,6) and (6 x 3, 6 x 10) as equivalent to (3,10). To compare (4,6) and (3,10), therefore, we compare (40,60) and (18,60).

Because 40 > 18, we say the family (40,60) > (40,18) and, consequently that the family (4,6) > (3,10). This, of course, harmonizes with our experience of these ideas applied to measurement – we want 'four - sixths' of a length to be greater than 'three - tenths' of it.

Notation

Traditionally in schools, rational numbers have been written as fractions and they have been described as 'one number over another' (the numerator and denominator). To be very precise it is not possible to have numbers over each other in the sense that one is on top of another. Numbers are not physical objects; they are ideas. They cannot be distributed in space. They only occur one after another in a time sequence.

The signs or symbols used to represent the numbers *can*, on the other hand, be arranged in space. We are correct, therefore, in saying that a fraction is the written sign consisting of a mark for one whole number on top of another mark for a whole number.

This may sound too precise, too much of a semantic quibble. It is not intended to do so. In fact, once the understanding is clear it hardly matters what words are used and we can, with clarity, continue to speak of 'numbers on top of other numbers'. It is a question of how best the underlying ideas can be communicated, suggesting that teachers exercise great care and sensitivity in recognizing the possibilities of different meanings arising within learners, as a result of words which can have many and quite different meanings.

In our discussion here we have used the ordered pair form of notation, not because we are trying to convey the basic ideas in a new light and we wish not to attract any prejudices a reader may have. Both the ordered pair and the fraction are valid and should be used interchangeably as convenience and context suggests, provided both forms arise from and are used for the meanings we have implied.

The ordered pair form (2,3) uses a comma to distinguish the symbol from (23) or possibly (2 + 3). The fraction ⅔ uses the line in lieu of the comma. On a typewriter ⅔ is seldom used; 2/3 is typed as a substitute. In some modern typographical fonts and highway signs $^2_3|$ is written for 'two thirds' with no comma or line to separate the two numerals 2 and 3.

In the foregoing we could just as correctly write:

"If $\frac{a}{b}$ and $\frac{c}{d}$ are pairs of the same rational number family then a x d = c x b.

$\frac{a}{b} > \frac{c}{d}$ if a > c (so long as b $\neq$ 0, d $\neq$ 0)"

... and so on.

Such forms are in most school text books. The ordered pair forms could and *should* be there just as commonly!

Mathematics

"*Definition:* The set R of rational numbers consists of all couples (a,b) of integers a and b $\neq$ 0. The "equality" of couples is governed by the convention that,

(a,b) = (a', b') if and only if ab' = a'b

This is the manner in which rational numbers are defined according to the

1953 edition of *A Survey of Modern Algebra* by Birkhoff and MacLane, a well established university text. We have used 'pairs' instead of 'couples'. The sign = used by them denotes 'equality for couples'. We have said 'equivalent', in that they are members of the same family. 'Integers' are the whole numbers, the negative numbers and zero, although zero is excluded in this definition; ab′ is equivalent to a x b′, a product.

A second definition indicates the slight differences in expression between some mathematicians, though they would accept each others when preceded by different previous development.

> "*Definition:* For each member (a,b) of T, R(a,b) is the set of all members (u,v) of T for which av = bu holds. We call each subset R(a,b) of T a *rational number.*"

NOTE: T symbolizes all ordered pairs (a,b), where a is an integer and b is an integer other than zero.

> Levi, Howard. *Elements of Algebra.* New York: Chelsea Publishing Co., 1953.

We have used 'family' instead of 'set' and implied that R(a,b) is a subset of T when we said earlier in the Summary: 'Each member is a pair of whole numbers, not *any* pair, but pairs such that certain connections . . . exist'.

Ratio and Proportion

Traditionally, the word *ratio* has been used to denote an ordered pair of whole numbers. One speaks of two lengths, for example, in the 'ratio of 8 to 10' and the written form 8:10 has been common. In this example we also say that the first length is 'eight - tenths' of the second and, reciprocally, the second is 'ten - eighths' of the first. Thus, the ratio of 8 to 10, the ordered pair (8,10) and the fraction $\frac{8}{10}$ are equivalent ways of representing the same relationship.

One can also have a ratio with more than two terms. 8:10:14:6 is a four - term ratio, symbolizing relationships between four lengths or sets of counters. A rational number can, therefore, be seen as a special - case ratio family, each member of which has only two terms.

The family (8:10:14:6) or written (8,10,14,6), has members:

$$(4,5,7,3)$$
$$(2 \text{ x } 4, 2 \text{ x } 5, 2 \text{ x } 7, 2 \text{ x } 3)$$
$$(3 \text{ x } 4, 3 \text{ x } 5, 3 \text{ x } 7, 3 \text{ x } 3) \dots \text{ and so on.}$$

Generally, if (a,b,c,d) and (u,v,w,x) are each sets of four whole numbers, they belong to the same ratio family, if and only if,

av = bu   from which it follows also that

aw = cu   bw = cv, bx = vd and cx = wd.

and ax = du

The standard name of a ratio family will be, as previously for a two - term ratio, that member whose terms have no common factor, other than 1.

The word *proportion* has also been used, traditionally, to identify two members of a rational number family. A proportion is consequently a statement that two ratios are equal.

The notation 4:5::8:10 is sometimes used. It is read as "4 and 5 are in the same ratio as 8 and 10", or in our nomenclature, "they are members of the same rational number family".

$$\frac{-6+6-1}{7 \quad 7 \quad 7}$$

$$+\frac{10}{7}+\frac{5}{7}+\frac{4}{7}$$

$$+\frac{9}{7}+\frac{2}{7}+\frac{9}{7}$$

$$\frac{10}{7}+\frac{10}{7}+\frac{9}{7}$$

$$\frac{15}{7}+$$

$$32\frac{-3}{7}$$

$$\frac{2\cdot4}{30}+\frac{2}{7}+2+1=\frac{29}{7}$$

$$\frac{20+9}{7}\quad\frac{39-10=\frac{29}{7}}{7}$$

$$\frac{16+13}{7 \quad 7}$$

$$\frac{12+13+4}{7 \quad 7 \quad 7}$$

$$\frac{18+11}{7 \quad 7}$$

$$\frac{1+9}{7 \quad 7}+\frac{2}{7}+\frac{8}{7}+\frac{9}{7}$$

$$10+10+10-10+5$$

$$\frac{31-2}{7 \quad 7}$$

$$\frac{2}{7}+\frac{9-1}{7 \quad 7}$$

$$10+10+1+1+3+1+1+1=27$$

13

Operations on Fractions, When Representing Rational Numbers

Having developed, by one means or another, the understanding that a rational number is:

> a family, each of whose members is a pair of whole numbers (or integers, if we wish to include negative numbers) such that there exists a certain connection between the four whole numbers of any two members of the family, viz. the product of the first of the first pair and the second of the second pair is equivalent to the product of the other two,

we are in a position to ask, "What can be *done* with such families which will be good mathematically and which can be applied to various problems in commerce, science and measurement?"

Two approaches seem open to us. We can try to invent or guess as to the combinations two such families can produce and then see where this leads. Alternatively we can play with sets of counters, rods, or other measuring devices to see whether we can discover something that will work satisfactorily with them.

For example, suppose we begin with the two families represented by (4,7) and (3,8). Could these be consistently related to another rational number family by using in some way, the four whole numbers involved: 4, 7, 3, 8? We might think of the family (7,15) because $4 + 3 = 7$ and $7 + 8 = 15$. This is a frequent guess by students, probably tempted understandably to do this by our conventions of writing:

$$\frac{4}{7} + \frac{3}{8} = \frac{7}{15}$$

His own examples

It can be argued that the student may not be helped much by being told he is wrong. For a kind of adding *has* been used, a fraction has been obtained as the "answer' and it does seem attractive to proceed as one reads, from left to right in horizontal lines across the page.

If the person guessing that $\frac{4}{7} + \frac{3}{8} = \frac{7}{15}$ also wants $\frac{4}{7}$ to be equivalent to $\frac{8}{14}$ then, providing that he wishes also to be consistent, he presumably would operate thus:

$$\frac{8}{14} + \frac{3}{8} = \frac{11}{22}$$

That implies he has to accept $(7,15)$ and $(11,22)$ as being members of the same rational number family! Guesses like this, therefore, produce contradictions when taken a little further. The student has to guess differently!

Again, someone's guess may suggest the addition of $(4,7)$ and $(3,8)$ produce 22, all four numbers being added together! But the pairs $(4,8)$ and $(3,7)$ on the same hypothesis would also lead to 22. So would $(3,4)$ and $(7,8)$ and presumably $(1,2), (1,2), (1,2)$ and $(12,1)$!

In this way a point of absurdity would soon be reached by anyone seriously suggesting that adding fractions is a process of adding all their whole number terms to produce the standard name of one number.

Many such inventions can be attempted with similar contradictions arrived at. A few more are suggested here:

$$\frac{4}{7} + \frac{3}{8} = \frac{11}{11} \dots \text{ because } 4 + 7 = 11 \text{ and } 3 + 8 = 11$$

This would seem to imply that:

$$\frac{4}{7} + \frac{3}{8} = \frac{4}{7} + \frac{8}{3}$$

and therefore, that $\frac{3}{8} = \frac{8}{3}$, contradicting the significance of the order of the whole number components of each pair. Besides, we shall not accept the common measuring interpretation that eight - thirds can be a substitute for three - eighths!

$\frac{4}{7} + \frac{3}{8} = \frac{7}{15}$ ... because $4 + 3 = 7$ and $7 + 8 = 15$

This would seem to suggest that,

$$\frac{3}{6} + \frac{1}{1} + \frac{3}{8} = \frac{7}{15}$$

and other names for $\frac{7}{15}$ would be

$$\frac{1}{2} + \frac{1}{2} + \frac{2}{3} + \frac{3}{8};$$

$$\frac{1}{2} + \frac{1}{2} + \frac{1}{3} + \frac{1}{2} + \frac{1}{2} + \frac{1}{2} + \frac{1}{2};$$

$$\frac{1}{1} + \frac{1}{1} + \frac{1}{1} + \frac{1}{1} + \frac{1}{1} + \frac{1}{1} + \frac{1}{9} \dots \text{ and so on.}$$

So, if anyone suggesting this, wants $\frac{1}{2}$ to be the symbol for our common understanding of 'one half' these are untenable names for 'seven fifteenths'. For $\frac{7}{15}$ is presumably less than 1 and the three examples above are decidedly greater than 1.

We come to the conclusion that such attempted combinations of two rational number families are absurd! (Which is *not* to imply that we have acted absurdly in proceeding thus.)

Using Rods

It is common sense to call a relationship between the length of a light - green rod and a pink rod, 'three fourths' and write either $(3, 4)$ or $\frac{3}{4}$. The green rod is three fourths of the pink length and that makes the order precise: green, pink.

The relationship between the lengths of the black and the pink, again takes its name from a combination of the two whole - number names involved, 'seven fourths' $(7, 4)$ or $\frac{7}{4}$. That of the relationship between the orange length and the pink is 'ten fourths', $(10, 4)$ or $\frac{10}{4}$.

It seems reasonable, therefore, to combine the $(7,4)$ relationship with that of the $(3,4)$ and associate with it the relationship $(10,4)$ because an orange length is equivalent to 'light - green + black'.

Moreover, it seems reasonable to write:

$$(3,4) + (7,4) = (10,4)$$

or

$$\frac{3}{4} + \frac{7}{4} = \frac{10}{4}$$

The sign $+$ in this context does not mean 'add together the two whole numbers represented on either side' (as it did formerly), because there are *four* whole numbers present. It means when the second component of each of the two pairs is the same, *then* we add the two first components.

If we write:

$$\frac{3}{4} + \frac{7}{4} = \frac{3+7}{4}$$

the meanings of the first $+$ sign and the second $+$ are different, the first signifying a combining of two *pairs* of whole numbers, the second a combination of two *single* whole numbers. It *is* possible to define a wider meaning for $+$ which embraces both these contexts, though for young learners the intended meaning will be satisfactory if they are securely aware of the context viz. what is being done with rational numbers.

So far, so good. Does this process, this invention, fit in with the rational number family notions? Let us investigate:

> If (3,4) added to (7,4) gives (10,4)
> then (6,8) added to (14,8) gives (20,8).
> (6,8) and (3,4) are equivalent
> So, too, are (7,4) and (14,8) and, indeed, (10,4) = (20,8)

On further investigation we find that any other two equivalents to (3,4) and (7,4) can be used in a similar process to result in a member of the (10,4) family. We are truly involved in combining *families* .

To prove that this can always be done, consider the families (a,b) and (c,d). According to the process under consideration the representatives (a,b) and (c,d) of these families cannot be combined, for b and d are *not* necessarily equal. Can we use appropriate equivalents? Yes, for generating the members of each family we get,

> (a,b) = (2a, 2b) = (3a, 3b) = . . . and so on
> (c,d) = (2c, 2d) = (3c, 3d) = . . . and so on

Is there a representative member of each family with the same second number? Either by trial and error, or by knowing about common multiples, we realize that:

> (a,b) = (d x a, d x b)
> (c,d) = (b x c, b x d)   (b x d = d x b)

So, instead of using (a,b) or $\frac{a}{b}$ and (c,d) or $\frac{c}{d}$, we are justified in substituting for them respectively (d x a, d x b) or $\frac{d \times a}{d \times b}$ and (b x c, b x d) or $\frac{b \times c}{b \times d}$.

Finally:$\frac{a}{b} + \frac{c}{d}$

$$= \frac{d \times a}{d \times b} + \frac{b \times c}{b \times d} = \frac{(d \times a) + (b \times c)}{d \times b} \text{ or} \frac{(d \times a) + (b \times c)}{b \times d}$$

This process which is called the 'addition of rational numbers' or the 'addition of fractions' applies, therefore, to all whole numbers represented by a, b, c, d, save only that neither b nor d can be zero, as has been previously decided.

This latter generalized awareness comes easily from rods. Consider any two pairs b rods or trains of rods. If the second train of each pair are the same, the process described above corresponds to adding together the first rod or train of each pair. This with the same second component of the first two pairs constitutes the new pair.

However, if we begin with (pink, yellow) and (red, light - green), the second components are *not* equivalent; yellow and light - green are not the same length.

These pairs give the relationship known as (4,5) and (2,3). Equivalents which are other members of the two families can now be set up. This occurs by a process of trial and error or by knowing that one seeks a common multiple – in this case one of 5 cm and 3 cm.

So to represent the first pair we choose (o + r, o + y) because o + r is the same length as 4 light - greens; and o + y is five light - greens. To represent the second pair (2,3), we choose (o, o + y).

The two pairs, equivalents of what we began with, now have a common second term. We, therefore, add the first two trains to get a length of '2 orange + red'. The final pair is (2 orange + red, orange + yellow), or expressed as a result of measuring both lengths in centimetres: (22,15).

In this manipulation we can see that both the 4 cm and 5 cm lengths were multiplied by 3; the 2 cm and 3 cm lengths were multiplied by 5.

The pair (22,15) is thus more revealingly written as

$$[\boxed{4 \times 3} + \boxed{5 \times 2}, 3 \times 5], \text{ or traditionally as fractions:}$$
$$\frac{4}{5} + \frac{2}{3} = \frac{(4 \times 3) + (5 \times 2)}{5 \times 3}$$

Now the rods are such that when a student has used them for a while he senses that what can be done with the four rods which represented (4,5) and (2,3) can be done with any other four rods or rod - trains, save only possible special cases involving 0 and 1. This has only to be tested in more cases to be substantiated, at least intuitively. In every instance it is seen to be inevitable that if two rational numbers with equal terms are combined as described,

$\frac{a}{b}$ and $\frac{c}{b}$, say, then $\frac{a+c}{b}$ is the 'result'. Moreover, beginning with ordered pairs as representatives of two families with *different* second terms, $\frac{d}{e}$ and $\frac{f}{g}$, say, then the result is another rational number family of which $\frac{(d \times g) + (e \times f)}{e \times g}$ is the immediately acceptable representative.

Mathematics

Mathematically this corresponds to the definition of one of the possible operations upon rational numbers:

"... while sums ... are defined by

$$(a, b) + (a', b') = (ab' + a'b, bb')"$$

This is a continuation of the previous quotation (chapter 12) from Birkhoff & MacLane's *Modern Algebra* .

We have seen a development which makes sense of this definition in that the implications can be applied to the measurement of lengths (or rods, in this case) and clearly also to sets of counters, pebbles or fingers.

Short Cuts

As a learner develops mastery in the operation just studied he will do well, of course, to consider whether in some cases short cuts are worthwhile. An obvious example is the addition of $\frac{1}{2}$ and $\frac{1}{2}$. Intuitively most students know this as 1 and it would be foolish to suggest the only way of finding the equivalent standard for $\frac{1}{2} + \frac{1}{2}$ is to apply the details of the process of generalization:

$$\frac{1}{2} + \frac{1}{2} = \frac{(1 \times 2) + (2 \times 1)}{2 \times 2} = \frac{2 \times 2}{2 \times 2} = \frac{4}{4} = \frac{1}{1} = 1$$

Again, given the pairs $\frac{12}{16}$ and $\frac{15}{27}$, for example, the student certainly can apply his understanding and skill to the generalized process:

$$\frac{12}{16} + \frac{15}{27} = \frac{(12 \times 27) + (16 \times 15)}{16 \times 27}$$

$$= \frac{324 + 240}{432}$$

$$= \frac{564}{432}$$

$$= \frac{141}{108}$$

$$= \frac{47}{36}, \text{ the standard form.}$$

However, if the two pairs are first 'reduced' to their own standard forms, the subsequent sequence of equivalents needs less arithmetic:

$$\frac{12}{16} + \frac{15}{27} = \frac{3}{4} + \frac{5}{9}$$

$$= \frac{(3 \times 9) + (4 \times 5)}{4 \times 9}$$

$$= \frac{47}{36}$$

Traditionally, the latter algorithm has been stressed early in fraction work, perhaps to the detriment of generalized solutions. Our thrust is to develop the understanding of the generalized form much earlier, mastering the skill of using it by having students make up and process many examples for themselves. That done, the students can be encouraged to discuss the value and place of short cuts, which can only be useful in some cases. It hardly seems worthwhile to spend much time looking for a short way when one can use the always - working pattern:

$$\frac{a}{b} + \frac{c}{d} = \frac{ad + bc}{bd}$$

NOTE:

Such short cuts and the process by which one transforms one family member into an equivalent, come traditionally under the heading of 'cancellation'. The word is not one to evade. We have avoided it until now and suggest that it too frequently carries the meaning of 'crossing out' as a skill for 'getting the answer'. It is the understanding which may suggest a crossing out as a way of noting on paper what corresponds in the mind, not the process of crossing out in itself, that is important. Given that approach, this is very acceptable:

$$\frac{\cancel{12}^{3}}{\cancel{16}_{4}} + \frac{\cancel{15}^{5}}{\cancel{27}_{9}}$$

Without the understanding, the marks on the paper can and often do cause great confusion.

Further Challenges

We can now consider adding three or more fractions, using rods, pebbles or just symbols.

Given: $\frac{3}{5} + \frac{4}{7} + \frac{6}{11}$ how shall we proceed to the standard name?

We certainly can find a standard substitute for

$\frac{3}{5} + \frac{4}{7}$, namely, $\frac{(3 \times 7) + (5 \times 4)}{5 \times 7}$.

The original name can, therefore, be transformed into:

$$\frac{(3 \times 7) + (5 \times 4)}{5 \times 7} + \frac{6}{11}$$

This represents a sum of two fractions, so we can *again* employ the general pattern, changing it to,

$$\frac{(3 \times 7) + (5 \times 4) \times 11 + 6 \times (5 \times 7)}{(5 \times 7) \times 11}$$

Can this be simplified?

$$6 \times (5 \times 7) = 6 \times 5 \times 7$$
$$(5 \times 7) \times 11 = 5 \times 7 \times 11$$
$$[(3 \times 7) + (5 \times 4)] \times 11 = [(3 \times 7) \times 11] + [(5 \times 4) \times 11]$$

by distributing the 'x 11' .

So $\frac{3}{5} + \frac{4}{7} + \frac{6}{11} = \frac{(3 \times 7 \times 11) + (4 \times 5 \times 11) + (6 \times 5 \times 7)}{(5 \times 7 \times 11)}$

A longer expression, of four fractions, say, may be easier in the sense that the general pattern is beginning to be mastered:

$$\frac{3}{5} + \frac{4}{7} + \frac{6}{11} + \frac{8}{13} = \frac{(3x7x11x13) + (4x5x11x13) + (6x5x7x13) + (8x5x7x1}{(5x7x11x13)}$$

If we wish to use letter names instead of specific number names, A for the number Mr. A is thinking of, B for Mr. B's number . . . and so on, then:

$$\frac{A}{B} + \frac{C}{D} + \frac{E}{F} + \frac{G}{H} =$$

$$\frac{(A \times D \times F \times H) + (C \times B \times F \times H) + (E \times B \times D \times H) + (G \times B \times D \times F)}{(B \times D \times F \times H)}$$

and if one can securely leave out the multiplications signs,

$$\frac{A}{B} + \frac{C}{D} + \frac{E}{F} + \frac{G}{H} = \frac{ADFH + CBFH + EBDH + GBDG}{BDFH}$$

We can note that the products in the numerator each consists of one of the fraction numerators and all the denominators of the *other* fractions. This is directly related to the fact that to equate all the denominators to the same number, the first fraction $\frac{A}{B}$ would have to be represented by $\frac{ADFH}{BDFH}$. B does *not* occur in the numerator product here, only DFH, because DFH is the *complement* of B in BDFH.

Subtraction

All the developments of the combination called 'subtraction of fractions' are similar to those just studied for 'addition', except that $+$ signs are changed to $-$ signs at the appropriate places.

$$\frac{7}{4}-\frac{3}{4}=\frac{7-3}{4}$$

$$\frac{3}{4}-\frac{7}{4}=\frac{3-7}{4} \quad \text{(even though the standard name for 3 - 7 might not}$$
be known at this stage)

$$\frac{4}{5}-\frac{2}{3}=\frac{(3 \times 4)-(5 \times 2)}{5 \times 3}$$

$$\frac{a}{b}-\frac{c}{b}=\frac{a-c}{b} \quad \text{(with no need to decide here whether a } > \text{c, a } < \text{c,}$$
or a $=$ c)

$$\frac{12}{16}-\frac{15}{27}=\frac{(12 \times 27)-(16 \times 15)}{(16 \times 27)}$$

$$\frac{3}{4}-\frac{5}{9}=\frac{(3 \times 9)-(4 \times 5)}{(4 \times 9)}$$

$$\frac{3}{5}+\frac{4}{7}-\frac{6}{11}=\frac{(3 \times 7 \times 11)+(4 \times 5 \times 11)-(6 \times 5 \times 7)}{5 \times 7 \times 11}$$

$$\frac{3}{5}-\frac{4}{7}-\frac{6}{11}=\frac{(3 \times 7 \times 11)-(4 \times 5 \times 11)-(6 \times 5 \times 7)}{5 \times 7 \times 11}$$

$$\frac{A}{B}\pm\frac{C}{D}\pm\frac{E}{F}\pm\frac{G}{H}=\frac{ADFH \pm CBFH \pm EBDH \pm GBDF}{BDFH}$$

If one wishes, one can study the subtraction process with rods and/or pebbles. Instead of two rods end - to - end to denote addition, they can be placed side - by - side for subtraction, order being significant. In the case of pebbles, some will be taken away to denote subtraction.

Mixed Fractions

One other matter remains in our study of addition and subtraction of 'fractions' as rational numbers, a relatively small matter but one which traditionally has loomed as very important in school instruction.

We say 'four and three sevenths' and write $4\frac{3}{7}$, a 'mixed fraction', because it apparently consists of a mixture of a whole number, 4, and a fraction $\frac{3}{7}$. To be precise, a whole number cannot be added to a family of which $(3, 7)$ is one representative, so mathematically, $4\frac{3}{7}$ would have to be interpreted as short for $\frac{4}{1} + \frac{3}{7}$. Nevertheless, in school there is a limit to the degree of purity and sophistication aimed at. What is more important is for students to:

a) have the awareness that $4\frac{3}{7}$ can be substituted by

$4 + \frac{3}{7}$, or $\frac{4}{1} + \frac{3}{7}$, and therefore, by $\frac{(4 \times 7) + (1 \times 3)}{1 \times 7}$.

$4\frac{3}{7}$ is accepted as equivalent, therefore, to $\frac{31}{7}$ and this *is* a fraction representing a rational number.

b) have the ability to transform any 'mixed' fractions to 'unmixed' equivalents and vice versa.
This will lead to other short cuts.
Given: $4\frac{3}{7} + 2\frac{5}{9}$ a learner *can* transform it into,

$$\frac{31}{7} + \frac{23}{9}$$

and thence to,

$$\frac{(31 \times 9) + (7 \times 23)}{7 \times 9}$$

or $\quad \dfrac{279 + 161}{63}$

or $\quad \dfrac{440}{63}$

and back into mixed fraction form as $6\frac{62}{63}$.

Alternatively, however, a learner can be encouraged to proceed thus:

$$4\frac{3}{7} + 2\frac{5}{9} = 6 + \frac{3}{7} + \frac{5}{9}$$
$$= 6 + \frac{27 + 35}{63}$$
$$= 6\frac{62}{63}$$

This may well raise his level of awarenes so that in future he prefers to adopt the shorter sequence of transformations.

Summary

We have parted considerably from the traditional emphasis and order of the presentation of the addition/subtraction of fractions, viz.

1. say what a fraction is by presenting a few examples of their written symbols, possibly with a few references to 'pies' .
2. hasten on to something called the 'addition of fractions' in which one is enjoined to go through certain processes on paper, cross out here and there, to get an answer that some authority external to ourselves says is correct.
3. deal with so - called easy fractions like $\frac{1}{2}, \frac{1}{4}, \frac{3}{8}$, first, and then trust that understanding will come so that years later the process, the 'algebra of the transformations' , will be mastered.

Our alternative includes,

a. developing the basic understanding of what rational numbers are, as described in the 'family notion' , that is, *algebraically* ,
b. study possibilities of combining pairs of families so that in every possible example each pair of families will correspond to a third family,
c. choose perhaps from many possibilities by requiring that what is finally chosen harmonizes with our common sense ways of adding lengths, or sets of counters,
d. proceed to master the structures by which the pairs of whole numbers are combined,
e. practise, inventing examples oneself, to master the process skills,
f. investigate cases where short cuts, legitimately made, save energy and time,
g. become aware of special cases and any restrictions on possibilities (denominator $\neq 0$),
h. decrease the emphasis on standard names, but seek non - standards which reveal the patterns of the structures,
i. have confidence that what one does is backed closely by sound mathematics, even if not yet in formal terms. Later work will not contradict, only make things more precise in technically accepted conventional terms,
j. know that the conventional algebraic treatments presented tradition-ally in high schools only involve possible changes of symbols and more complications.

e.g. $\frac{a-b}{a+b} + \frac{a \sin A}{b \cos A}$ can be processed similarly to $\frac{A}{B} + \frac{C}{D}$.

The 'new' symbols in numerators and denominators do *not* affect the step to,

$$\frac{(a-b) \times b \cos A + (a+b) \times a \sin A}{(a+b) b \cos A}$$

Later, after considering other operations on rational numbers we shall suggest another radical change, that the addition of fractions is not necessarily the first operation to be stressed for children, but that multiplication may qualify for that role.

Multiplication and Division of Rational Numbers

Primarily, a question as before can be asked. Can two rational number families be combined in such a way that with each two another family can be associated, the process by which the combination is established being the same for all representative members of the families? This is what we did in the previous section. We found two ways of so combining and we finally called them the addition and subtraction of rational numbers. The meanings ascribed to the words 'addition' and 'subtraction' were extensions of our previous understandings of the same words as used with whole numbers.

The question now challenges us to find other ways of forming combinations.

One approach arises from the need to identify particular members of a particular rational number family. Given the family (1,4) for example, we know that members are,

$$(1,4), (2,8), (3,12) \ldots \text{and so on.}$$

Rather than always stringing out the members – or some of them – to locate one particular member, we could ask "In the (1,4) family what is the first number (or term) of the pair whose second term is 8?"

We write, $(1, 4)$ of $8 = 2$, or $\frac{1}{4}$ of $8 = 2$

Immediately we are able to set up a sequence, triggered by this:

$\frac{1}{4}$ of $12 = 3$ because $(1, 4) = (3, 12)$

$\frac{1}{4}$ of $16 = 4$ because $(1, 4) = (4, 16) \ldots$ and so on.

Alternatively, we may choose any rectangle in the traditional multiplication table display, excluding the zero row and column, and consider the four entries at the corners of the rectangle.

1	2	3	4	5	6
2	4	6	8	10	12
3	6	9	12	15	18
4	8	12	16	20	24

The rectangle chosen gives as its corners,

1 . . 4
2 . . 8

Beginning at one of these entries and moving clockwise or counter-clockwise around the others, we can extract:

<table>
<tr><td colspan="2">Clockwise</td><td colspan="2">Counter-clockwise</td></tr>
</table>

Clockwise		*Counter-clockwise*	
$(1,4)$ of $8 = 2$ (or $\frac{1}{4}$ of $8 = 2$)		$\frac{1}{2}$ of $8 = 4$ (or $(1,2)$ of $8 = 4$)	
$(4,8)$ of $2 = 1$		$\frac{2}{8}$ of $4 = 1$	
$(8,2)$ of $1 = 4$			
$(2,1)$ of $4 = 8$		$\frac{8}{4}$ of $1 = 2$	
		$\frac{4}{1}$ of $2 = 8$	

Counters

Such statements harmonize with what we get from counters. 'One half' of 8 counters is 4 counters. (or 'one twoth of 8 is 4, since 'one twoth' indicates there are *two* of them in 1').

$\frac{2}{8}$ of 4 with counters may seem puzzling unless one is a master of equivalences of fractions. Then one knows that, $\frac{2}{8} = \frac{1}{4}$ and so $\frac{2}{8}$ of $4 = 1$.

$\frac{8}{4}$ of $1 = \frac{2}{1}$ of 1. This makes unusual reading, perhaps ("Two oneths of one"?), but $\frac{2}{1}$ can be interpreted as 2 and '2 of 1' as 2. This rather stilted language used in a special case does not contradict the patterns already established and is therefore acceptable.

Lengths

Similarly, if one resorts to lengths, a brown rod is equivalent to 8 whites. 'One half of a brown' must be a pink length, and pink is in length equivalent to 4 whites. Again $\frac{1}{2}$ of $8 = 4$ and all the other statements made above can be used with similar meanings gleaned from manipulation of rods.

However, we do not yet have two rational number families combined in a new way. An extension of what we have just done may suggest a further step.

Focussing on the Language

If 'one half of 8 is 4' , what would be 'one half of 8 *tenths*'? Surely, '4 *tenths*', analogous to,

$\frac{1}{2}$ of 8 *metres* is 4 *metres*

$\frac{1}{2}$ of 8 *grams* is 4 *grams*

$\frac{1}{2}$ of 8 *apples* is 4 *apples*

In 'one half of 8 tenths = 4 tenths' we *do* seem to be combining two families to be associated with a third; (1,2) represents the first, (8,10) the second, (4,10) the third. We can test whether this applies equally well to other members of each family by substitutions:

$$\frac{1}{2} \text{ of } \frac{8}{10} = \frac{1}{2} \text{ of } \frac{4}{5} = \frac{2}{5}, \text{ a member of the (4, 10) family}$$

$$\frac{1}{2} \text{ of } \frac{8}{10} = \frac{1}{2} \text{ of } \frac{16}{20} = \frac{8}{20}, \text{ a member of the (4, 10) family} \ldots \text{ and so on.}$$

If '*one* half of 8 tenths' is 4 tenths, then '*two* halves of 8 tenths' must surely be '*two* times four', or 8, tenths.

$$\frac{2}{2} \text{ of } \frac{8}{10} = \frac{2 \times 4}{10}, \text{ and}$$

$$\frac{3}{2} \text{ of } \frac{8}{10} = \frac{3 \times 4}{10}$$

$$\frac{4}{2} \text{ of } \frac{8}{10} = \frac{4 \times 4}{10}$$

$$\frac{5}{2} \text{ of } \frac{8}{10} = \frac{5 \times 4}{10} \ldots \text{ and so on.}$$

We can now use substitutes for *any* of the fractions in $\frac{1}{2}$ of $\frac{8}{10} = \frac{4}{10}$:

$$\frac{2}{4} \text{ of } \frac{8}{10} = \frac{4}{10}$$

$$\frac{10}{20} \text{ of } \frac{8}{10} = \frac{4}{10}$$

$$\frac{10}{20} \text{ of } \frac{80}{100} = 10 \times \frac{1}{20} \text{ of 80 hundredths} = 10 \times \frac{4}{100} = \frac{40}{100} = \frac{4}{10}$$

There are endless possibilities.

All this suggests what we seek: another effective combination of two families. Can the process apply to *any* two families, for instance, $\frac{3}{4}$ and $\frac{5}{9}$? In every example used above, we see that the first denominator was a factor of the second numerator. In $\frac{3}{4}$ of $\frac{5}{9}$ however, 4 is *not* a factor of 5. With the availability of an endless number of equivalents of $\frac{3}{4}$ and of $\frac{5}{9}$ however we transform conveniently, choosing a common multiple of 4 and 5; 20 will do:

$$\frac{3}{4} \text{ of } \frac{5}{9}$$

$$= \frac{3}{4} \text{ of } \frac{20}{36} = \frac{15}{35} \text{ (by the process developed above)}$$

Fractions as Operators on Whole Numbers

An alternative approach is to consider $\frac{3}{4}$ of $\frac{5}{9}$ as a combination of two operators, asking whether there is one fractional operator which has an equivalent effect.

$\frac{5}{9}$ of 36, as we have seen, is 20

so $\frac{3}{4}$ of $\left(\frac{5}{9}\text{ of }36\right) = \frac{3}{4}$ of $20 = 15$

BUT, $15 = \frac{15}{36}$ of 36

So, $\frac{3}{4}$ of $\frac{5}{9}$ operating upon 36 is equivalent to $\frac{15}{36}$ operating on 36,

We write: $\frac{3}{4}$ of $\frac{5}{9} = \frac{15}{36}$

or $\frac{3}{4}$ of $\frac{5}{9} = \frac{3 \times 5}{4 \times 9}$

Generally for whole numbers a, b, c, d (b, d $\neq$ 0),
$$\frac{a}{b}\text{ of }\frac{c}{d} = \frac{a \times c}{b \times d}$$

$$\frac{2}{9} \times \frac{9}{7} = \frac{2}{7}$$

$$\frac{2}{11} \times \frac{11}{8} = \frac{2}{8}$$

$$\frac{18}{65} \times \frac{65}{112} \times \frac{112}{42} = \frac{18}{42}$$

$$\frac{24}{7} \times \frac{7}{8} \times \frac{8}{64} = \frac{24}{64}$$

$$\frac{9}{100} \times \frac{100}{99} \times \frac{99}{4} \times \frac{4}{3} = \frac{9}{3}$$

$$\frac{4}{3} \times \frac{3}{6} \times \frac{6}{99} = \frac{4}{99}$$

$$\frac{17}{1,000} \times \frac{1,000}{2} \times \frac{2}{108} = \frac{17}{108}$$

$$\frac{20}{4} \times \frac{4}{5} \times \frac{5}{2} = \frac{20}{2}$$

$$\frac{14}{102} \times \frac{102}{9} \times \frac{9}{1} = \frac{14}{1}$$

$$\frac{7}{2} \times \frac{2}{9} = \frac{7}{9}$$

Multiplication of fractions

This is called the 'multiplication of fractions' or perhaps, the 'multiplication of rational numbers expressed as fractions'. It can be remarked that in a sense the fractions are *not* multiplied. The multiplication is of the whole numbers which are parts of the fractions. The pair of numerator whole numbers gives one product, the denominators the other.

Counters and/or Rods

Operators can be used on counters as just previously on whole numbers. As before, a suitable set has to be chosen. If 36 counters are selected, a subset equal to five - ninths gives 20 counters, i.e., five of the subsets of which nine in union give the 36 counters.

Three quarters of the 20 counters is 15.

So, once again, 'three quarters of the five - ninths of 36 counters is 15 counters'.

$$\frac{3}{4}\,of\frac{5}{9} = \frac{15}{36}$$

With this statement as a beginning, many other true sentences can be generated, especially in this example because 36 is rich in factors,

$$\left(\frac{1}{2}\,of\,\frac{2}{18}\right)\,of\,36 = \frac{1}{18}\,of\,36 \qquad \frac{1}{2}\,of\,\frac{2}{18} = \frac{1}{18}$$

$$\left(\frac{2}{2}\,of\,\frac{2}{18}\right)\,of\,36 = \frac{2}{18}\,of\,36 \qquad \frac{2}{2}\,of\,\frac{2}{18} = \frac{2}{18}$$

$$\left(\frac{3}{2}\,of\,\frac{2}{18}\right)\,of\,36 = \frac{3}{18}\,of\,36 \qquad \frac{3}{2}\,of\,\frac{2}{18} = \frac{3}{18}$$

$$\left(\frac{4}{2}\,of\,\frac{2}{18}\right)\,of\,36 = \frac{4}{18}\,of\,36 \qquad \frac{4}{2}\,of\,\frac{2}{18} = \frac{4}{18}$$

... and so on   ... and so on

$$\frac{1}{3}\,of\,\frac{3}{18} = \frac{1}{18} \qquad\qquad \frac{1}{4}\,of\,\frac{4}{18} = \frac{1}{18}$$

$$\frac{2}{3}\,of\,\frac{3}{18} = \frac{2}{18} \qquad\qquad \frac{2}{4}\,of\,\frac{4}{18} = \frac{2}{18}$$

$$\frac{3}{3}\,of\,\frac{3}{18} = \frac{3}{18} \qquad\qquad \frac{3}{4}\,of\,\frac{4}{18} = \frac{3}{18}$$

$$\frac{4}{3}\,of\,\frac{3}{18} = \frac{4}{18} \quad \text{... and so on.} \qquad \frac{4}{4}\,of\,\frac{4}{18} = \frac{4}{18} \quad \text{... and so on}$$

It would seem generally that $\frac{A}{B}$ of $\frac{B}{C}=\frac{A}{C}$. When the first denominator is equivalent to the second numerator the single associated family is that which can be represented by the first numerator and the second denominator. The repetition of the 'B' is ignored in the answer; 'B' is cancelled!

We have, it should be noted, said nothing explicitly about, for example $\frac{3}{4}$ of $\frac{5}{9}$ of 37, or any other number of which 36 is *not* a factor. It seems intuitively sensible, however, to expect $\frac{3}{4}$ of $\frac{5}{9}$ of 37 counters to be equivalent to $\frac{15}{36}$ of 37 counters, although this cannot easily be shown directly using 37 counters.

Using Rods

Elementary examples of these compounded relationships can arise as soon as a student is familiar with the fractional names given to a pair of rods.
    Black is *seven* whites, so a white is *one - seventh* of a black.

White is *one sixth* of a dark green,
    one fifth (? five - th) of a yellow
    one fourth of a pink
    one third (? one three - th) of a light - green
    one half (? one two - th) of a red
    one (? one one - th) of a white
    one eighth of a brown

    . . . and so on.

Consider the relationships between the lengths of three rods: red, yellow, black for instance.

red $=\frac{2}{7}$ of the black length

yellow $=\frac{5}{7}$ of the black length

red $=\frac{2}{5}$ of the yellow length

So, red $=\frac{2}{5}$ of $\left(\frac{5}{7}\text{ of the black length}\right)$

There are two names, therefore, for the relationship (red, black), 'two sevenths' and 'two fifths of five sevenths',

$$\frac{2}{7} = \frac{2}{5}\text{ of }\frac{5}{7}$$

A few more examples convince that given any two fractions such that the first denominator is the same as the second numerator, then when they are combined in this way, an equivalent is given by the first numerator and the second denominator.

A further challenge, providing practice and greater mastery, could be to use *four* lengths: red, yellow, black, and orange, say:

$$\text{red} = \frac{2}{5} \text{ yellow} \qquad \text{yellow} = \frac{5}{7} \text{ black} \qquad \text{black} = \frac{7}{10} \text{ orange}$$

But red is also $\frac{2}{10}$ orange.

So, red $= \frac{2}{10}$ orange $= \frac{2}{5}$ of $\frac{5}{7}$ of $\frac{7}{10}$ of orange

We write, $\frac{2}{10} = \frac{2}{5}$ of $\frac{5}{7}$ of $\frac{7}{10}$, because both $\frac{2}{10}$ and $\frac{2}{5}$ of $\frac{5}{7}$ of $\frac{7}{10}$ operating on the orange length identify the red length.

The 'equivalent name game' can be played further by stressing more names for $\frac{2}{10}$ (or any other fraction) of the form just developed. One such selection from an infinity of possibilities is,

$\frac{2}{7}$ of $\frac{7}{10}$ $\frac{2}{7}$ of $\frac{7}{8}$ of $\frac{8}{10}$ $\frac{2}{a}$ of $\frac{a}{b}$ of $\frac{b}{c}$ of $\frac{c}{d}$ of $\frac{d}{10}$

$\frac{2}{9}$ of $\frac{9}{10}$ $\frac{2}{9}$ of $\frac{9}{11}$ of $\frac{11}{10}$ ... and so on

$\frac{2}{23}$ of $\frac{23}{10}$ $\frac{2}{A}$ of $\frac{A}{B}$ of $\frac{B}{10}$

$\frac{2}{723}$ of $\frac{723}{10}$ ... and so on

$\frac{2}{A}$ of $\frac{A}{10}$

          ... and so on.

By substitution for any of the fractions the same generalised pattern is established as previously, regardless of the particular numbers of the representatives first presented.

Given $\frac{5}{9}$ of $\frac{10}{11}$ we transform each fraction because $9 \neq 10$:

$$\frac{5}{9} = \frac{50}{90}, \quad \frac{10}{11} = \frac{90}{99}$$

So, $\frac{5}{9}$ of $\frac{10}{11} = \frac{50}{90}$ of $\frac{90}{99} = \frac{50}{99}$

In general, given four whole numbers a, b, c, d, (b, d $\neq$ 0),

$$\frac{a}{b} \text{ of } \frac{c}{d} = \frac{a \times c}{b \times c} \text{ of } \frac{b \times c}{b \times d} = \frac{a \times c}{b \times d}$$

Does 'of' Mean Multiplication?

The process just studied is *not* precisely the same kind of multiplication which combined two whole numbers. It implies, in the new context, two such whole number products, one of a and c, one of b and d.

The term 'multiplication' is used again probably for three reasons:

1. The new process can reasonably be called a 'kind of' multiplication.
2. In the case of special rational numbers, for example,

   $\frac{2}{1}$ is used as a symbol for $2 \div 1 = 2, \frac{3}{1} = 3, \frac{6}{1} = 6.$

   There is, therefore, a very close connection between the new process for fractions whose denominators are 1, and the 'old' multiplication.
3. In the mathematical development of sets, if there are two operations, both of which uphold the commutative and the associative principles, and if one of them is distributive over the other, then the first is called 'multiplication', the other 'addition'.

Commutative, Associative and Distributive Principles

Commutative: $\frac{a}{b}$ of $\frac{c}{d} = \frac{a \times c}{b \times d}$

$$= \frac{c \times a}{d \times b} \quad \text{(multiplication of whole numbers } is$$
commutative)

$$= \frac{c}{d} \text{ of } \frac{a}{b}$$

So this operation for fractions is also commutative.

Associative: $\frac{a}{b}$ of $\left(\frac{c}{d} \text{ of } \frac{e}{f} \right)$

$$= \frac{a}{b} \text{ of } \frac{c \times e}{d \times f}$$

$$= \frac{a \times (c \times e)}{b \times (d \times f)}$$

$$= \frac{(a \times c) \times e}{(b \times d) \times f} \quad \text{(Multiplication of whole numbers } is \text{ asso-}$$
ciative).

$$= \left(\frac{a \times c}{b \times d} \right) \text{ of } \frac{e}{f}$$

$$= \left(\frac{a}{b} \text{ of } \frac{c}{d} \right) \text{ of } \frac{e}{f}$$

So this operation for fractions is also associative.

Distributive: We need to show that:

$$\frac{a}{b} \text{ of } \left(\frac{c}{d} + \frac{e}{f}\right) = \left(\frac{a}{b} \text{ of } \frac{c}{d}\right) + \left(\frac{a}{b} \text{ of } \frac{e}{f}\right)$$

$$\frac{a}{b} \text{ of } \left(\frac{c}{d} + \frac{e}{f}\right) = \frac{a}{b} \text{ of } \frac{cf + de}{df}$$

$$= \frac{a \times (cf + de)}{b \times df}$$

$$= \frac{acf + ade}{bdf}$$

(multiplication of whole numbers *is* distributive)

$$= \frac{acf}{bdf} + \frac{ade}{bdf}$$

$$= \frac{ac}{bd} + \frac{ade}{bdf}$$

$$= \frac{ac}{bd} + \frac{ae}{bf}$$

$$= \left(\frac{a}{b} \text{ of } \frac{c}{d}\right) + \left(\frac{a}{b} \text{ of } \frac{e}{f}\right)$$

This operation for fractions is distributive over addition.
It may be of interest to note that addition is *not* distributive over multiplication, or that, for example,

$$\frac{2}{3} + \left(\frac{4}{5} \text{ of } \frac{6}{7}\right) \neq \left(\frac{2}{3} + \frac{4}{5}\right) \text{ of } \left(\frac{2}{3} + \frac{6}{7}\right)$$

(For variation, instead of using a, b, c . . . etc., we have chosen 2, 3, 4, . . . etc. Generality is *not* lost provided we make no special assumptions about the use of these numbers rather than the use of any others).

$$\frac{2}{3} + \left(\frac{4}{5} \text{ of } \frac{6}{7}\right) \qquad = \frac{2}{3} + \frac{4 \times 6}{5 \times 7}$$

$$= \frac{(2 \times 5 \times 7) + (3 \times 4 \times 6)}{3 \times 5 \times 7}$$

$$\left(\frac{2}{3} + \frac{4}{5}\right) \text{ of } \left(\frac{2}{3} + \frac{6}{7}\right) \quad = \left(\frac{(2 \times 5) + (3 \times 4)}{3 \times 5}\right) \text{ of } \left(\frac{(2 \times 7) + (3 \times 6)}{3 \times 7}\right)$$

$$= \frac{[(2 \times 5) + (3 \times 4)] \times [(2 \times 7) + (3 \times 6)]}{3 \times 5 \times 3 \times 7}$$

$$= \frac{2 \times 2}{3 \times 3} + \frac{4 \times 2}{5 \times 3} + \frac{4 \times 6}{5 \times 7} + \frac{2 \times 6}{3 \times 7}$$

$$= \tfrac{2}{3} \left(\tfrac{2}{3} + \tfrac{4}{5} + \tfrac{6}{7} \right) + \tfrac{4 \times 6}{5 \times 7}$$

Unless, therefore, the sum of the original three factors $\tfrac{2}{3} + \tfrac{4}{5} + \tfrac{6}{7}$ happens to be 1, and in this example it is certainly not, the two expressions examined are not equivalent.

In general,

$$\tfrac{a}{b} + \left(\tfrac{c}{d} \text{ of } \tfrac{e}{f} \right) \neq \left(\tfrac{a}{b} + \tfrac{c}{d} \right) \text{ of } \left(\tfrac{a}{b} + \tfrac{e}{f} \right)$$

The above work is perhaps not necessary for school students but there is frequently a problem as to why, when fractions are multiplied, "one gets a smaller answer'. Previously, multiplication increased the numbers! The reason is that the meaning of the word 'multiplication' has probably changed – 'probably' because it depends upon the students' development. Many children note this by calling the operations with fractions, 'kind of addition' and 'kind of multiplication', the new meanings arising from what is done in the new context. This is no more strange, really, than the meaning of other words, though renewed discussion of this may be appropriate. The word 'father', for example, may to a young toddler refer to one specific man. Later he will use the same word to refer to many men, each one of which is involved in a relationship with others. Later still, the meaning of 'father' can be extended even more, but the old meanings will not be contradicted. It is the context which will guide a precise meaning.

Summary

Operations can be defined which each time associate two rational number families with a third family. Those called the 'addition and subtraction of fractions' are two such operations.

Another is to associate, for example, the family $\tfrac{2}{3}$ and the family $\tfrac{5}{8}$ with the family $\tfrac{10}{24}$. More generally, if a, b, c, d are any four whole numbers except that neither b nor d is zero, then the third family is $\tfrac{a \times c}{b \times d}$.

In developing and studying the implications of this, we used phrases like, 'two thirds of counters' or 'two thirds of a length'. We attached meaning to longer phrases like 'two thirds of five eighths of a set of counters'; each of these being found to be equivalent to a fraction, the denominator and numerator of which was a pair of products. Although *any* pile of counters or length of rods was not suitable for such discovery, whenever such piles or lengths *were* convenient, then, again, $\tfrac{a}{b}$ of $\tfrac{c}{d} = \tfrac{a \times c}{b \times d}$.

This operation is called 'multiplication' because it involves multiplication, and the commutative, associative and distributive principles are true.

Cancellation – Uses and Abuses

Any name of the form $\frac{a}{b}$ of $\frac{c}{d}$, or $\frac{a}{b} \times \frac{c}{d}$, is, as we have seen, equivalent to three other transformations of the same form using the same four whole numbers:

$$\frac{a}{b} \times \frac{c}{d} \qquad\qquad \frac{c}{d} \times \frac{a}{b}$$

$$\frac{a}{d} \times \frac{c}{b} \qquad\qquad \frac{c}{b} \times \frac{a}{d}$$

In some specific cases this suggests a useful technique. Consider, for example, $\frac{3}{16} \times \frac{4}{9}$. This can be transformed into $\frac{3}{9} \times \frac{4}{16}$ and immediately we note that:

$$\frac{3}{9} = \frac{1}{3}; \qquad \frac{4}{16} = \frac{1}{4}$$

So, $\quad \frac{3}{9} \times \frac{4}{16} = \frac{1}{3} \times \frac{1}{4} = \frac{1}{12}.$

-giving the standard form $\frac{1}{12}$.

Although in this example the saving of work hardly makes it worthwhile as a short - cut procedure, there may be great advantage for samples containing large numbers:

Compare, $\dfrac{144}{65} \times \dfrac{13}{9} = \dfrac{144}{9} \times \dfrac{13}{65} = \dfrac{16}{1} \times \dfrac{1}{5} = \dfrac{16}{5}$

with, $\dfrac{144}{65} \times \dfrac{13}{9} = \dfrac{144 \times 13}{65 \times 9} = \dfrac{1872}{585}$

... and now search for common factors of 1872 and 585!

Traditionally, this has been called 'cancellation' and as we have insisted previously in other circumstances, this is quite satisfactory. However, if what the students have done has established the impression that 'cancellation' means crossing out to get the right answer, then there are too many pitfalls for security in what turns out to rest on insufficient understanding. It is easy to 'cross out'. What is more difficult is to know exactly what one can legitimately cross out and what not.

Without adequate understanding, a person conditioned to cancelling by crossing out one numeral exactly beneath another may find it difficult to cancel if the fraction is written like this, $\frac{12}{9}$, the denominator not being exactly beneath the numerator!

Others can fall into the mechanical habit of crossing out diagonally:

$$\frac{144}{13\,\cancel{65}} \times \frac{\cancel{8}^{\,1}}{9}$$

without displaying as much ease with,

$$\frac{\cancel{5}^{1}}{\cancel{65}^{13}}$$

and yet this is at the heart of the equivalence concept.

Difficulties can also arise from a surfeit of cancellations with resulting illegibility, causing errors in subsequent reading,

$$\frac{\cancel{144}^{\cancel{48}^{16}}}{\cancel{65}^{13}} \times \frac{\cancel{5}^{1}}{\cancel{9}\cancel{3}_{1}}$$

All such traditional hazards tend to be avoided in practice if students have mastered the equivalence of fractions as rational numbers and what can be done with them.

Division of Fractions

Still another operation is possible. Perhaps the awareness of the 'multiplication' operation suggests the possibility of the inverse operation 'division', just as there is a division and multiplication of whole numbers.

From the inverse aspect we can ask,

'If $\frac{2}{3} \times \frac{4}{9} = \frac{2 \times 4}{3 \times 9}$, can we begin with $\frac{2 \times 4}{3 \times 9}$ and $\frac{4}{9}$ and produce $\frac{2}{3}$?

This is analogous to $4 \times 9 = 36$, $36 \div 9 = 4$,

Clearly $\frac{2 \times 4 \div 4}{3 \times 9 \div 9} = \frac{2}{3}$, so we can say:

$$\frac{2 \times 4}{3 \times 9} \div \frac{4}{9} = \frac{2}{3}$$

using $\div$ as a sign for this operation, division, inverse to multiplication.

Transformations of each fraction pair lead to equivalent statements:

$$\left(\frac{2 \times 4}{3 \times 9} \div \frac{4}{9}\right) \to \left(\frac{8}{27} \div \frac{4}{9}\right) \to \left(\frac{16}{54} \div \frac{4}{9}\right) \to \left(\frac{32}{108} \div \frac{4}{9}\right) \ldots \text{ and so on.}$$

$$\left(\frac{8}{27} \div \frac{8}{18}\right) \to \left(\frac{8}{27} \div \frac{16}{36}\right) \ldots \text{ and so on.}$$

$$\left(\frac{16}{54} \div \frac{12}{27}\right) \to \left(\frac{32}{108} \div \frac{36}{81}\right) \ldots \text{ and so on.}$$

Given now $\frac{16}{11} \div \frac{5}{9}$, for example, we note that 16 is not a multiple of 5, 11 is not a multiple of 9. We could however proceed thus:

$$\frac{16}{11} \div \frac{5}{9} = \frac{3\frac{1}{5}}{1\frac{2}{9}} = \frac{3\frac{1}{5} \times 5}{1\frac{2}{9} \times 5} = \frac{16}{5\frac{10}{9}} = \frac{16 \times 9}{5\frac{10}{9} \times 9} = \frac{144}{55}$$

Alternatively, we can substitute suitable equivalents for $\frac{16}{11}$

$\frac{16 \times 5 \times 9}{11 \times 5 \times 9}$ will be convenient.

$$\frac{16}{11} \div \frac{5}{9} = \frac{16 \times 5 \times 9}{11 \times 5 \times 9} \div \frac{5}{9} = \frac{16 \times 9}{11 \times 5} = \frac{144}{55}.$$

BUT $\frac{16 \times 9}{11 \times 5} = \frac{16}{11} \times \frac{9}{5}$, so it appears that

$$\frac{16}{11} \div \frac{5}{9} = \frac{16}{11} \times \frac{9}{5}$$

and we conclude that an equivalent of 'dividing by $\frac{5}{9}$' is 'multiplying by $\frac{9}{5}$'. $\frac{9}{5}$ is the reciprocal of $\frac{5}{9}$.

We have not lost generality by using the example we chose, although in more traditional algebraic form,

$$\frac{a}{b} \div \frac{c}{d} = \frac{a \times c \times d}{b \times c \times d} \div \frac{c}{d} = \frac{a \times d}{b \times c} = \frac{a}{b} \times \frac{d}{c}.$$

Traditionally, it is the 'multiplication by the reciprocal' which has been taught by rote methods. While this *is* one of the ways of looking at the situation we need to study from the learning point of view which development is psychologically more attractive. If one has remembered that the multiplication of fractions implies the separate products of the given numerators and denominators, a learner may well think that for the new division operation, numerators and denominators can be divided. This, as we see, is also valid and can always be accomplished by suitable substitutions.

Counters and Rods

Given two sets of counters, one of say 11 counters, the other of 5, each bears a relationship to another set, of 23 counters, say. The relationships are expressed by $\frac{11}{23}$ and $\frac{5}{23}$.

In reply to the question, "What is 11 counters divided by 5 counters?" the answer of $2\frac{1}{5}$ can be given.

Extending this to the relationship names $\frac{11}{23}$ and $\frac{5}{23}$ we can say $\frac{11}{23} \div \frac{5}{23}$ is also $2\frac{1}{5}$. The fact that two sets of counters can be measured by another set does not change the relationship $2\frac{1}{5}$.

Any other set could just as easily be chosen as the set to which our first two sets of counters are related.

SO $\dfrac{11}{24} \div \dfrac{5}{24} = 2\dfrac{1}{5}$

$\dfrac{11}{25} \div \dfrac{5}{25} = 2\dfrac{1}{5}$

$\dfrac{11}{x} \div \dfrac{5}{x} = 2\dfrac{1}{5}$ . . . and so on.

Similarly with the use of rods. The stimulus, "Divide an orange + white length by yellow" causes one to construct a train of yellow rods alongside the orange + white length and report what happens. One such report is that $11 \div 5 = 2\dfrac{1}{5}$.

But both the orange + white and a yellow length can, in their turn, be measured by any length; orange + orange + light - green is one. Once again, therefore, we can say

$\dfrac{11}{23} \div \dfrac{5}{23} = 2\dfrac{1}{5}$

$\dfrac{11}{24} \div \dfrac{5}{24} = 2\dfrac{1}{5}$ (with 0 + 0 + p as the unit)

. . . and so on.

The generalized form follows as before.

Commutativity, Associativity & Distributivity with Division

The principles can be studied in the context of division.

Commutativity $\dfrac{a}{b} \div \dfrac{c}{d}$ is not generally equivalent to $\dfrac{c}{d} \div \dfrac{a}{b}$, for

$\dfrac{a}{b} \div \dfrac{c}{d} = \dfrac{a \div c}{b \div d}$ and $a \div c \neq c \div a, \ b \div d \neq d \div b.$

Division of rational numbers is not commutative.

Associativity $\left(\dfrac{a}{b} \div \dfrac{c}{d}\right) \div \dfrac{e}{f} = \dfrac{ad}{bc} \div \dfrac{e}{f} = \dfrac{adf}{bce}$ Generally these are not equivalent because

$\dfrac{a}{b} \div \left(\dfrac{c}{d} \div \dfrac{e}{f}\right) = \dfrac{a}{b} \div \dfrac{cf}{dc} = \dfrac{ade}{bcf}$ $\dfrac{f}{e} \neq \dfrac{e}{f}$

Division of rational numbers is not associative.

Distributivity   For distributivity we need to examine whether,

$\dfrac{a}{b} \div \left(\dfrac{c}{d} + \dfrac{e}{f}\right) = \left(\dfrac{a}{b} \div \dfrac{c}{d}\right) + \left(\dfrac{a}{b} \div \dfrac{e}{f}\right)$

Now $\dfrac{a}{b} \div \left(\dfrac{c}{d} + \dfrac{e}{f}\right) = \dfrac{a}{b} \div \left(\dfrac{cf + de}{df}\right)$

$$= \frac{adf}{b(cf + de)} \qquad \ldots (i)$$

And $\left(\dfrac{a}{b} \div \dfrac{c}{d}\right) + \left(\dfrac{a}{b} \div \dfrac{e}{f}\right) = \dfrac{ad}{bc} + \dfrac{af}{be}$

$$= \frac{abde + afbc}{bcbe} \qquad \ldots (ii)$$

Generally these two fraction expressions (i) and (ii) are not equivalent, so division is *not* distributive over addition.

Summary

The traditional difficulties of understanding the division of fractions is transformed into common sense approaches — either abstract approaches using only spoken and written symbols, or others using concrete aids such as counters or colored rods.

In each approach a procedure is developed for relating two rational number families to a third such that there is an inverse relationship to that which we called multiplication, involving the same three families.

Connections Between Whole Numbers and Rational Numbers
Given a, b, c, d, any whole number (or integer), b, d $\neq$ 0, then

$$\frac{a}{b} \times \frac{c}{d} = \frac{a \times c}{b \times d}$$

$$\frac{a}{b} \div \frac{c}{d} = \frac{a \div c}{b \div d} = \frac{a \times d}{b \times c}$$

Special cases of these operations are interesting. If the denominators are 1, we have

$$\frac{a}{1} \times \frac{c}{1} = \frac{a \times c}{1}$$

$$\frac{a}{1} \div \frac{c}{1} = \frac{a \div c}{1} = \frac{a}{c}$$

These have strong correspondences with the statements about whole numbers a, c, their product a x c and their quotient a ÷ c, frequently written $\dfrac{a}{c}$.

Thus, there is a subset of the rational numbers (those rationals with 1 as the second term of the standard name) which corresponds to whole numbers and the operations +, =, x, ÷.

Whole Numbers	Rational Numbers
$2, 3, \ldots 9 \ldots 87 \ldots$	$\dfrac{2}{1}, \dfrac{3}{1}, \ldots \dfrac{9}{1} \ldots \dfrac{87}{1} \ldots$
$2 \times 3 = 6$	$\dfrac{2}{1} \times \dfrac{3}{1} = \dfrac{6}{1}$
$6 \div 3 = 2$	$\dfrac{6}{1} \div \dfrac{3}{1} = \dfrac{2}{1}$
$4 + 7 = 11$	$\dfrac{4}{1} + \dfrac{7}{1} = \dfrac{11}{1}$
$7 - 4 = 3$	$\dfrac{7}{1} - \dfrac{4}{1} = \dfrac{3}{1}$
$4 - 7 = {}^-3$	$\dfrac{4}{1} - \dfrac{7}{1} = \dfrac{{}^-3}{1}$ $\ldots$ and so on.

In this sense the whole number, (or the integer) system can be viewed as a subset of the rationals. There is a close correspondence between the two systems. The integer system has been extended with deeper meanings to the operational words, add, subtract, multiply and divide. In technical terms an *isomorphism* exists between the two systems.

Referring once more to Birkhoff and MacLane (p. 115),

> "... specifically (isomorphism) is a one - one correspondence between F and F' such that if $x \leftrightarrow x'$ and $y \leftrightarrow y'$ then, $(x + y) \leftrightarrow (x' + y')$ and $(xy) \leftrightarrow (x'y')$.'

In this case F is the set of integers, F' the rational numbers; x and y are any members of F, x' and y' members of F'.

Because of this close connection the same symbol is often used for an integer and its corresponding rational. It is not possible, except in context, to know whether $2 + 3$ say, is representing an integer or a rational. In both cases the standard name is 5 so that no difficulty arises provided one is aware of the number system of the particular context.

14

Decimals – Number Names Using Dots

The essence of the decimal system of notation is that it provides another way of representing rational numbers. Decimals are new *numerals* , not new numbers, numerals which are not fractions.

The fraction $\frac{1}{10}$ and therefore, the rational number for which it is a representative, is re-named 'point one', or 'decimal one' and written .1.

Immediately, therefore, $\frac{2}{20}$ is also .1 and so are $\frac{3}{30}, \frac{4}{40}, \frac{5}{50} \ldots$ etc.

$\frac{2}{10}$ will be .2, $\frac{3}{10} = .3$, $\frac{4}{10} = .4 \ldots \frac{9}{10} = .9$

It may at first be thought that $\frac{10}{10}$, according to this schema should be written .10 ('point ten'). However, $\frac{10}{10} = 1 + \frac{0}{10}$ so a preferable way is 1.0. (Later we shall write .01 for $\frac{1}{100}$, so .10 will then be $\frac{1}{10} + \frac{0}{100}$, and that is equivalent to .1. We cannot continue with the ambiguity that .10 be 1 or $\frac{1}{10}$, so we are obliged to have $\frac{10}{10}$ as 1.0)

$$\frac{14}{10} = 1 + \frac{4}{10} = 1 + .4 = 1.4$$

Many equivalents of the new form are now available:

$\frac{25}{10} = 2.5$, $\frac{37}{10} = 3.7$, $\frac{89}{10} = 8.9$, $\frac{893}{10} = 89.3 \ldots$ and so on.

Each of these can yield other equivalents which may appear to sophisticates as trivial but mastery of which is necessary for later work: e.g., 3.7 = 03.7, 003.7 = 0003.7 . . . and so on, and perhaps before specifically saying so: 3.70, 3.700, 3.7000 . . .

Historical Note

The small dot constituting the decimal point is deceptively important. It looks so insignificant yet its presence and precise location becomes very significant indeed. Some appreciation of this can be gained by studying the extension of the place value system. The numeral 763 is read '7 hundred 6ty 3', the 7 being in the hundreds column, the 6 in the tens column, the 3 in the ones column. If there was another digit following the 3, 7634 say, what would the 4 probably represent?

The relation suggested by the sequence 100, 10, 1 is that of a 'one tenth' transformation from each name to the next as we look from left to right. The new digit, therefore, would represent four 'one tenths'.

76345 would imply '5 tenths of tenths' or '5 hundredths' . . . and so on.

The question then arises that, presented with 76345, how would a reader know that the 7, for example, would represent 7 hundred and not 7 hundred thousand? Some sign is necessary to avoid the ambiguity.

At various times in history different ways were used and many of them can be re-invented for the purpose of study now:

763/45   763 45
                        H
763-45   763₄₅   76345

Different colored digits could also be used for the 763 as compared to 45.

We are, these days, left with the convention 763.45 in North America, 763· 45 in Europe.

Addition and Subtraction

As soon as a decimal form is established a previously used strategy can be employed again. Responding to, "What are other names for .4?" they include:

.3 + .1	.5 - .1	2 x .2	0.4	0.40
.2 + .2	.6 - .2	4 x .1	00.4	0.400
.2 + .1 + .1	7 - .3	8 x ½ x .1	000.4	0.4000

. . . and so on   . . . and so on   . . . and so on   . . . and so on

If only 'tenths', or 'the first decimal place', is involved, there is hardly much else to consider for the addition and subtraction of decimals. Given, say, .4 + 1.6 + .7 + 0.9, one can either rewrite it 'vertically' which conveniently places the 'tenths' , or the decimal places, in line.

Alternatively one can proceed to get the standard name for the sum without so doing:

$$
\begin{aligned}
&\begin{array}{l}
.4 \\
1.6 \\
.7 \\
\underline{0.9}
\end{array}
\end{aligned}
\qquad
\begin{aligned}
&= .4 + 1.6 + .7 + 0.9 \\
&= 2.0 + .7 + 0.9 \\
&= 2.7 + 0.9 \\
&= 3.6
\end{aligned}
$$

One could also revert to fraction forms:

$$
\begin{aligned}
&= .4 + 1.6 + .7 + 0.9 \\
&= \frac{4}{10} + 1\frac{6}{10} + \frac{7}{10} + \frac{9}{10} \\
&= 1\frac{26}{10} \\
&= 3\frac{6}{10} = 3.6
\end{aligned}
$$

All the intermediate steps are not necessary, and every student should practice the forms most convenient to himself.

For subtraction, all the previously studied patterns still hold, so that successive transformations can be used to arrive at the standard name – if that is the objective.

$$
\begin{aligned}
&36.5 - 19.8 \\
&= 36.7 - 20.0 \\
&\doteq 16.7
\end{aligned}
\qquad
\begin{aligned}
&36.5 - 19.8 \\
&= 17.5 - 0.8 \\
&= 17.7 - 1.0 \\
&= 16.7
\end{aligned}
\qquad
\begin{aligned}
&36.5 - 19.8 \\
&= 36.0 - 19.3 \\
&= 35.9 - 19.2 \\
&= 16.7 \\
&\ldots \text{and so on.}
\end{aligned}
$$

Whether introduced soon after the .1 form, or left until some additions and subtractions have been practiced, the form for $\frac{1}{100}$ will be .01. This suggests the pattern, $\frac{1}{1000} = .001$, $\frac{1}{10000} = .0001$ ...and so on, the number of zeros in the decimal being one less than those in the denominator.

$$\frac{4}{100} \text{ will be } .04$$

$$\frac{17}{100} \text{ will be } .17$$

$$\frac{98}{100} \text{ will be } .98$$

$$\frac{100}{100} \text{ will be } 1 + \frac{0}{100} \text{ or } 1 + \frac{0}{10} + \frac{0}{100} = 1.00$$

Appropriate addition and subtraction strategies follow as before. They correspond to the addition of whole numbers except for the important placement of the decimal point:

2 8010	2801.0	280.10	280.1
6900	690.0	69.00	69
152437	15243.7	1524.37	1524.37

Given:
$$2.93 + 6.1 + .02 + 19 + .96$$

it may be helpful first to use the equivalent,
$$2.93 + 6.10 + .02 + 19.00 + .96$$

each numeral being of the same form with two decimal places. This initial transformation lessens the subsequent miswriting of terms.

The extension of the scheme to thousandths, and further, follows similarly.

Simple introductions of decimal forms can be begun with primary children by naming a white rod as .1 of an orange. With that, the red length becomes .2 or .1 + .1 (of an orange) and so on. Some temporary ambiguity may be met concerning the orange length but by moving on to orange + yellow, say, as 1 + .5 or 1.5, the orange + white length has to be 1.1 and the orange 1.0 is seen to be a new written name for 1.

Multiplication

'Four times .1' is clearly .4, either from the relationship of four whites to an orange length, or four counters measured by a set of ten counters.

Other products, with their corresponding standards follow:

$$4 \times .2 = .8 \qquad 40 \times .1 = (10 \times 4) \times .1 = 10 \times .4 = 4$$
$$4 \times .3 = 1 + .2 = 1.2 \qquad 40 \times .2 = 8$$
$$4 \times .4 = 1.6 \qquad 40 \times .3 = 12$$
$$\dots \text{ and so on} \qquad \dots \text{ and so on}$$

It is not possible ever to anticipate which sequences will be developed by particular learners but many are possible. Using those already obtained above,

$$4.0 \times .2 = .8 \qquad 40.0 \times .1 = 4.0$$
$$4.0 \times .3 = 1.2 \qquad 40.0 \times .2 = 8.0$$
$$4.0 \times .4 = 1.6 \qquad 40.0 \times .3 = 12.0$$
$$\dots \text{ and so on}$$

To discuss the process of getting the standard corresponding to *any* product we need only be aware of what the equivalent process would entail if the rationals were in fraction form.

$$\text{e.g., } 4.6 \times 17.28 = \frac{46}{10} \times \frac{1728}{100}$$

$$= \frac{46 \times 1728}{10 \times 100}$$

$$= \frac{46 \times 1728}{1000} = \frac{79488}{1000} = 79.488$$

Alternatively, therefore, we can temporarily omit the decimal points, find the standard for the product as if all terms were whole numbers, then finally adjust by dividing, in this case by 1000. This amounts to inserting the point to allow for three decimal places.

$\left(\frac{1}{10}\right.$ corresponds to <u>one</u> place, $\frac{1}{100}$ to <u>two</u>, $\frac{1}{1000}$ to <u>three</u> ... and so on.$\left.\right)$

That there are *three* decimal places in the above example comes from the sum of the decimal places in the terms of the product, one from 4.6, two from 17.28. This sum of decimal places will correspond to the powers of ten in the denominators of the standard *fraction* forms and then to the number of decimal places in the final standard *decimal* form.

Generally:

$$\left(\begin{array}{c}\text{a numeral with} \\ \text{x decimal places}\end{array}\right) \times \left(\begin{array}{c}\text{a numeral with} \\ \text{y decimal places}\end{array}\right) = \left(\begin{array}{c}\text{standard name with} \\ (x + y) \text{ decimal places}\end{array}\right)$$

Or, in fraction form:

$$\frac{\begin{array}{c}\text{numeral with zero}\\\text{decimal places}\end{array}}{10^x} \times \frac{\begin{array}{c}\text{numeral with zero}\\\text{decimal places}\end{array}}{10^y} = \frac{\begin{array}{c}\text{standard corres-}\\\text{ponding to product}\end{array}}{10^{x+y}}$$

Summary

The traditional rule for multiplying decimals is valid: "multiply the two numerals as though there were no decimal points present. Insert the point in the standard form according to the pattern: the number of places is equal to the sum of the separate decimal places of the numbers in the original product."

Another strategy is to estimate the standard name by using approximations of each term in a product. 2.74 x 7.6, for example, will be greater than 2 x 7 and less than 3 x 8, i.e., between 14 and 24. We can multiply 274 by 76 therefore, obtain 20824 and insert the point to show 20.824, the only possibility between 14 and 24.

If the terms are less than 1 this strategy may not be so easy.   .274 x .076 will be greater than .2 x .07 and less than .3 x .08 or between .014 and .024, but one needs perhaps more confidence to use this check with such numbers. Nevertheless, of course, the strategy can still be used.

A combination of the various developments is worthwhile because of the ever present probability of error in placing the point. Large amounts of money have been lost on account of a missing or misplaced point!

Division

For the division of decimals the approach can be similar, namely, develop sequences beginning with an example, preferably known to be correct.

If an orange rod length is considered the unit, white is .1 of it. Blue is .9, so .9 ÷ .1 = 9. The length of a brown is .8 so:

$$.8 \div .1 = 8$$
$$.7 \div .1 = 7 \quad \text{(using black)}$$
$$1.8 \div .1 = 18 \quad \text{(using orange + brown)}$$
$$\ldots \text{and so on}$$

If a ten-orange length (or 1 metre) is the unit, white becomes .01, blue is .09 and we have,

$$.09 \div .01 = 9$$
$$.08 \div .01 = 8$$
$$.18 \div .01 = 18 \quad \text{(orange + brown is now .18)}$$
$$\ldots \text{and so on}$$

In such examples the number of decimal points in the terms of the quotient are the same. There is correspondence, of course, to

$$\frac{9}{10} \div \frac{1}{10} = 9, \frac{9}{100} \div \frac{1}{100} = 9 \text{ etc.}$$

If the number of decimal points of one term is different from that of another term, we get

$$.9 \div .01 = 90 \qquad \text{(.9 corresponds to 9 blues, .01 to 1 white)}$$
$$9 \div .01 = 900$$
$$90 \div .01 = 9000$$

and $.09 \div .001 = 90$ (100 orange is 1, .001 is white, .09 is .090 is 9 oranges)

$.09 \div .0001 = 900$ (1000 orange is 1, .0001 is white, .09 is 90 oranges)

$$\ldots \text{and so on}$$

Alternatively, just as in multiplication we can *add* the decimal places in the product terms, for division we can *subtract* the places in quotient terms:

.32 ÷ .1 gives (2 - 1) decimal places in the standard form: 3.2;

.32 ÷ .01 gives (2 - 2) or 0 places; standard is 32;

.032 ÷ .01 gives (3 - 2) or 1 place; standard is 3.2;

3.2 ÷ .01 gives (1 - 2) or ⁻1 place.

If 3.2 shows 1 place of decimals, 32. shows 0 places, then evidently 32 . shows (1 - 2) or 'negative one' place, though if a space is left it may be recognized otherwise. A place holder of a zero is, therefore, used, 320. or just 320 without a decimal point. If we say 'the decimal point moves' we mean, of course, that it has *appeared* to have moved.

So much for any quotient which corresponds to a standard name with zero 'remainder'. In other words, when the second term of a quotient is a *factor* of the first. What of an example like 28.742 ÷ 4.9? 49 is *not* a factor of 28472. Equivalent substitutes are again at the heart of effective strategy as we see in the next section.

Division of Decimals as Repeated Subtraction

We could proceed directly, using the notion that division is a form of repeated subtraction:

$$4.9\overline{)28.472}(5$$
$$\underline{24.5}$$
$$3.972 \qquad \longrightarrow \qquad 5\frac{3.972}{4.9}$$

All these answers are equivalent. However, we can also write the fraction in decimal form.

Alternatively the quotient could be transformed into 284.72 ÷ 49 or into 2847.2 ÷ 490 or into 28472 ÷ 4900.

$$49\overline{)284.72}(5$$
$$\underline{245}$$
$$39.72 \qquad \longrightarrow \qquad 5\frac{39.72}{49} \text{ which, of course is equivalent to}$$

$$5\frac{3.972}{4.9}$$

$$490\overline{)2847}(5$$
$$\underline{2450}$$
$$397.2 \qquad \longrightarrow \qquad 5\frac{397.2}{490}$$

$$4900\overline{)28472}(5$$
$$\underline{24500}$$
$$3972 \qquad \longrightarrow \qquad 5\frac{3972}{4900}$$

Decimal Equivalents of Any Fraction

Consider, for the moment, the simple fraction $\frac{3}{7}$ before returning to the division of decimals.

If $\frac{3}{7}$ can be transformed into a fraction with denominator 10, the numerator provides the numeral for the decimal form. There is no pair of whole numbers in the (3,7) family, however, whose second component is 10. However, we can say that,

$$\frac{3}{7} = \frac{4\frac{2}{7}}{10} \text{ because } 3 \times 10 = 7 \times 4\frac{2}{7}$$

Therefore $\frac{3}{7}$ can be written $.4\frac{2}{7}$

Alternatively,

$$\frac{3}{7} = \frac{\frac{3}{7}}{1} = \frac{100 \times \frac{3}{7}}{100 \times 1} = \frac{\frac{300}{7}}{100} = \frac{42\frac{6}{7}}{100}$$

So $\frac{3}{7} = .42\frac{6}{7}$

Proceeding similarly, $\frac{3}{7} = \frac{428\frac{4}{7}}{1000} = .428\frac{4}{7}$

$$\frac{3}{7} = \frac{4285\frac{5}{7}}{1000} = .4285\frac{5}{7} \qquad \qquad \text{...and so on.}$$

This process can be used with any fraction. In particular, we can deal with the $5\frac{3972}{4900}$ previously used. (last section)

We know that the decimal will be of the form 5. and, maybe, have some approximate notion that $\frac{3972}{4900}$ is greater than .5, perhaps approximately .8. We can, however, go further because,

$$\frac{3972}{4900} = \frac{\frac{3972}{4900} \times 10}{10} = \frac{\frac{3972}{4900} \times 100}{100} = \frac{\frac{3972}{4900} \times 1000}{1000} \text{...and so on}$$

The numerators of each of these equivalents enables us to process a quotient.

$$4900\overline{)39720} \quad \text{or} \quad 4900\overline{)397200} \quad \text{or} \quad 4900\overline{)3972000} \dots \text{etc. --A}$$

$$\text{or} \ 8\frac{520}{4900} \qquad\qquad \text{or} \ 81\frac{300}{4900} \qquad\qquad \text{or} \ 810\frac{3000}{4900} \qquad \dots \text{and so on}$$

$$\text{giving} \ .8\frac{520}{4900} \qquad\qquad \text{or} \ .81\frac{300}{4900} \qquad\qquad \text{or} \ .810\frac{3000}{4900} \qquad \dots \text{and so on}$$

$$\text{Thus } 28.472 \div 4.9 \ = \quad 5.8\frac{520}{4900}$$

$$= \quad 5.81\frac{300}{4900}$$

$$= \quad 5.810\frac{3000}{4900} \ \dots \text{and so on}$$

In many applications the final fraction 'remainder' is omitted. This results in an approximately true statement. As a convention it is agreed that if this fraction is equal or greater than $\frac{1}{2}$ it is counted as 1, if $<\frac{1}{2}$ as zero.

$$28.472 \div 4.9 \simeq 5.8$$
$$\simeq 5.81 \quad (\simeq \text{ approximately equals})$$
$$\simeq 5.811$$

Summary

In practice there is no need to deal with many equivalents as we have done in detail above. Either one temporarily omits the decimal points and proceeds to process a quotient as though only whole numbers were involved, extending the numerator with as many zeros as are needed (see A in previous section); or, alternatively, one changes to an equivalent with perhaps only the denominator as a whole number. The usual process of division is then performed, the point being inserted in the standard form as it is reached.

In both cases approximations are wise, done either before or after the detailed computation, in order that the position of the decimal point be correct.

Another well known relationship is important here too, namely, the equivalence between a multiplication sentence and a corresponding one in division:

If a x b = c then c ÷ b = a, and vice versa.

If we assert, therefore, that $\dfrac{28.472}{4.9}$ gives an equivalent of 5.811, either exactly, or approximately, we must be aware that we are also asserting that 5.811 x 4.9 = 28.472. Thus, if we are not so confident about the rightness of our division, we can check it by our confidence in multiplication!

Different Bases

All the above discussion is similarly valid for decimal forms in bases other than the common ten. The differences are only in the names used. These can be obtained directly as developed in chapter 3 or extracted from a multiplication table.

For instance, on the 'pink planet', the base is ⟨xx / xx⟩ and the multiplication table is:

x	1	2	3	10
1	1	2	3	10
2	2	10	12	20
3	3	12	21	30
10	10	20	30	100

Hence 2 x 3 = 12, 12 ÷ 3 = 2 etc.

   2 x .3 = 1.2,   1.2 ÷ .3 = 2 etc.

$\dfrac{1}{10}$ will still be written as .1 and from this it follows that,

$$\frac{1}{2} = \frac{2}{10} = .2$$

$$\frac{1}{3} = \frac{1\frac{1}{3}}{10} = .1\frac{1}{3}$$

$$= \frac{11\frac{1}{3}}{100} = .11\frac{1}{3}$$

$$= \frac{111\frac{1}{3}}{1000} = .111\frac{1}{3} \ \ldots \text{and so on}$$

Only the digits 0, 1, 2, 3 are legitimately used in number names for the given base, although, of course, any of them are repeatable.

15

Per Dozens
Per Kilometres
Per Cents

$\frac{1}{2} = 6\frac{1}{2}$ per Bakers dozen

$\frac{1}{3} = 4\frac{1}{3}$ per bakers dozen

$\frac{1}{5} = \frac{13}{5}$ per bakers dozen

$\frac{1}{7} = \frac{13}{7}$ per bakers dozen

$\frac{1}{8} = \frac{13}{8}$ per bakers dozen

$\frac{1}{2} = 7$ per (bakers dozen + 1)

$\frac{1}{3} = \frac{14}{3}$ per (bakers dozen + 1)

$\frac{1}{2} = 7\frac{1}{2}$ per (bd + 2)

$\frac{1}{3} = 5$ per (bd + 2)

$\frac{1}{2} = \frac{1}{4}$ per $\frac{1}{2}$

$\frac{1}{2} = \frac{1}{8}$ per $\frac{1}{4}$

$\frac{1}{2} = \frac{1}{16}$ per $\frac{1}{8}$

$\frac{1}{2} = \frac{2}{2}$ per fly wings

$\frac{1}{2} = 50$ per century

$\frac{1}{4} = 25$ per c

$\frac{1}{5} = 20$ per c

$\frac{1}{20} = 5$ per c

$\frac{1}{10} = 10\ \%$

$\frac{1}{5} = 20\ \%$

$\frac{1}{7} = \frac{100}{7}\ \%$

Per dozens

Percentages are not new numbers. They are the same numbers as those represented by decimals and by fractions, namely, rational numbers. Percentages are names for rational numbers particularly convenient in certain contexts.

If we once again consider, for example, the family represented by $\frac{3}{5}$ we recognize that it contains an infinite number of members. If there is one member whose second whole number is 100, then the first of the pair is called the *percentage*. For $\frac{3}{5}$ the pair (60,100) is the significant member. It is said, therefore, that 'three fifths is 60 percent', written $\frac{3}{5} = 60\%$.

Traditionally, in schools percentage like so many other topics, has too frequently been dealt with in isolation. Teachers would be well advised, however, to have students consider the meanings of common phrases like,

3 minutes per hour
4 per dozen
8 hours per day

These can lead to equivalent fractions: '4 per dozen', for everyone who knows that a dozen is 12, gives one - third, so

$$4 \text{ per dozen} = \frac{1}{3}$$

$$3 \text{ per dozen} = \frac{1}{4}$$

$$6 \text{ per dozen} = \frac{1}{2}$$

Once an entry of this kind has been made students will very likely produce other examples. Not only this, they also suggest or accept abbreviations of phrases like 'per dozen', pd for instance. This may seem a small issue in itself but it provides another opportunity for students to develop their own symbolism without inhibition.

$$5 \text{ per score} = \frac{1}{4}$$

$$10 \text{ per score} = \frac{1}{2}$$

$$15 \text{ per score} = \frac{3}{4}$$

$$1 \text{ per baker's dozen} = \frac{1}{13}$$

$$2 \text{ per baker's dozen} = \frac{2}{13}$$

$$3 \text{ per baker's dozen} = \frac{3}{13} \qquad \text{... and so on.}$$

$\frac{1}{10} = 2 \text{ per score}$ $8 \text{ oz. per lb.} = \frac{1}{2}$

$\frac{2}{10} = 4 \text{ p.s.}$ $1 \text{ oz. per lb.} = \frac{1}{16}$

$\frac{3}{10} = 6 \text{ p.s.}$ $2 \text{ oz. per lb.} = \frac{2}{16} = \frac{1}{8}$

        ... and so on   ... and so on

In the cases where denominators of the fractions are not factors of the given number there is further challenge.

What is, for example, $\frac{1}{5}$ in 'per dozens'? In the case of $\frac{1}{3} = 4$ pd, $\frac{1}{2} = 6$ pd, there are relationships to 3 x 4 = 12, to 4 x 3 = 12, and to 2 x 6 = 12. For $\frac{1}{5}$ therefore, we can ask 5 x __ = 12, and reply 5 x $\frac{12}{5}$ = 12,

So, $\qquad \frac{1}{5} = 2\frac{2}{5}$ pd

$\frac{2}{5} = 2\frac{2}{5}$ x 2 pd = $4\frac{4}{5}$ pd

$\frac{3}{5} = 2\frac{2}{5}$ x 3 pd = $7\frac{1}{5}$ pd

$\frac{n}{5} = 2\frac{2}{5}$ x n   pd

Percentages

Development of this kind, the students inventing examples of their own, not only gives more practice in fraction manipulation, it produces percentages as a particular instance of a generalized principle.

For $\qquad \frac{1}{5} = 20$ per century

$\frac{2}{5} = 40$ per century

$\frac{3}{5} = 60$ p.c.

$\frac{1}{7} = 14\frac{2}{7}$ p.c. .... and so on.

All that is now needed is the information that p.c. is not commonly used outside the classroom and that for p.c. the conventional percentage sign % is substituted:

$$\frac{1}{7} = 14\frac{2}{7}\%$$

An interesting feature for many students may be a challenge to the teacher's repertoire by studying, for example, 'per year' or 'per leap year' or other new examples invented for the purpose. Opportunity can be taken for almost any number to be considered in this way and if one likes humorous examples the students may like to accept the teacher's age as 99 and deal with 'per teacher's age'!

All problems of percentages can, of course, be transformed into corresponding problems in fractions or decimals, although for greater efficiency in producing standard forms students need to study more direct and short routes.

Given the problem, 'what is $7\frac{1}{2}\%$ of \$24?, one *can* proceed thus:

$$7\frac{1}{2}\% = \frac{7\frac{1}{2}}{100} = \frac{15}{200} = \frac{3}{40}$$

$$\frac{3}{40} \text{ of } 24 = \frac{3}{5} \text{ of } 3 = \frac{9}{5} = 1\frac{4}{5} = 1.8$$

So $7\frac{1}{2}\%$ of \$24 $= \$1.80$.

But one should also aim at something like this:

$$7\frac{1}{2}\% \text{ of } \$24 \qquad 7\frac{1}{2} \times 24 = 168 + 12$$

$$= 180$$

Answer \$1.80

... and *not* written!

Difficulty in the past with percentages has been aggravated by a lack of appreciation that they are only different forms of names for rational numbers. If a class of students has been working well with our notions here, that rational numbers are families of ordered pairs of whole numbers, an interesting test of their mastery may be to see whether they master percentages in a very short time. For only *one* piece of information is needed, extra to a knowledge of rational numbers, namely that, for example, 37% stands for $\frac{37}{100}$, 37 per hundred or (37,100). *Everything else* has been done, no matter what commercial or measurement applications are then looked at in problems.

16

Exponents
Small Number Names Up High

Number names which are said to be in 'exponent form' can be introduced and used perhaps informally earlier than the position in this chapter suggests. They were mentioned in passing in chapter 1 although they could arise during a first study of multiplication.

When students come up with products like:

$$10 \times 10 \times 10 \times 10 \times 10 \times 10 \qquad 2 \times 2 \times 2 \times 2 \times 2 \qquad 5 \times 5 \times 5$$

a short form may be welcomed, triggered by a simple remark to the effect that new number names can be written for each, namely,

$$10^6 \qquad\qquad 2^5 \qquad\qquad 5^3$$

The names are read as '10 to the 6th power', '2 to the 5th power' and '5 to the 3rd power'. There is no necessity to introduce the more sophisticated sayings '5 cubed' or '4 squared' (4^2), though they may not be detrimental. It is just that 'cubed' may not remind one of 'three', or 'squared' of 'two'. The number written on the line in the standard exponential form shows what is the factor repeated in the product, the one written slightly smaller and raised 'up in the air' gives the number of terms.

Such introduction does *not* imply that a study of exponents has to follow immediately. It does, however, facilitate the use of energy-saving forms as a substitute for very long products and the mastery of such notation will be of great help in future mathematics.

When such a study is begun an entry can be made like that described above or from actions on concrete materials.

With Fingers

It is possible, of course, for students to share their fingers for a model of, say, 3 x 3 x 3. This needing '3 times three groups each of 3 fingers', however, at least three people will need to cooperate and is cumbersome anyway. 2 x 2 x 2 is easier in that one person has sufficient digits, but because it is the only example apart from the special 1 x 1 x 1 one can do on one's hands the introduction of exponents from manipulating fingers is very limited.

With Counters, Leading to Place Value

With a large supply of counters on the table the task becomes more manageable than with fingers. One way of doing this is by playing 'clumps', though this is not a conventional name.

The leader holds up a number of fingers and calls "clumps". The challenge for the players is to push the counters around into as many clumps as possible, each clump having as many counters as the number of fingers shown. This finished, the result may be a number of clumps with some single counters extra, not enough to make up another clump.

A clump can and should vary from one number to another. There is certainly nothing wrong in the leader saying 'four' or 'six', the usual names, but the generalized objective is to form clumps regardless of the particular size.

The preliminary part of the game mastered, a new phase occurs when the call comes, "clumps of clumps". One 'clump of clumps' is as many clumps as there are fingers in a clump within the same game. If a clump is of four fingers, a 'clump of clumps' will be a pile of 4 lots of 4 counters.

In general, the original set of counters is now subdivided into so many clumps of clumps, so many clumps and a few singles, not enough to make one more clump. Depending on the size of the original set and the particular size of clump chosen, there may be 'zero clumps of clumps', and/or zero clumps, and/or zero singles.

The next stage is to get 'clumps of clumps of clumps', though unless the original set is large enough there may not be any such formations possible. The formations get dramatically larger very quickly. This experience is very valuable, however, in the process, which is called 'exponentiation'.

Written notation can be used to record the results. Precise counting of each formation is not really important to get the sense of the process at first, though as soon as the counting can be done without inhibition accuracy can be welcomed.

e.g. Primitive recordings may include the like of:

3 clumps of clumps
2 clumps
and 4 singles   (clump:)

Later it can be shortened to: 3 cc, 2c, 4   or to: 4, 2c, 3cc.

At this juncture the exponent forms can be introduced,

$$3c^2, \quad 2c, \quad 4$$

If the set of counters is much larger than that reported here c^3, c^4, c^5, etc. may be included, although in a sense they can appear in recordings of small sets: $3c^2$, 2c, 4, $0c^3$, $0c^4$, $0c^5$ . . . and so on.

Place Value

Exponentiation is the essence of the place value system of writing numerals. It is clear from a recording such as '4, $3c^2$, 2c' which part refers to 'clumps of clumps', which to 'clumps' and which to singles. Conventionally, a clump is called 'ten' (and it always *can* be in some base, on *'some* planet') so the record of the number of counters becomes '4, 3 tens of tens, 2 tens', or '4, 3 hundreds, 2 tens'.

Historically, as is well known, economy was accomplished by omitting the writings of 'hundreds' and 'tens' and 'singles' the final word, 324, being used on the understanding that the righthand digit always refers to 'singles' the next to it to the left refers to tens, the next hundreds . . . and so on. In fact that is how it is read: "3 hundred 2 tens four". If such an arrangement were not understood there would be ambiguity as to which digit referred to 'tens', which to singles, etc.

Further, if one of the formations of counters *finally* does not exist, a zero has to be recorded in the appropriate place. 4032, for example, would be interpreted as '2 singles', '3 tens', 'zero tens of tens' and '4 thousands'. The conventional pattern is shown by:

tens of tens of tens thousands	tens of tens hundreds	tens	single ones
10^3	10^2	10	1

With Rods, Squares and Cubes

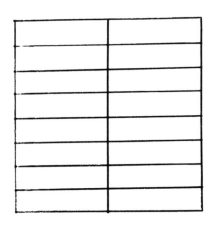

4 x 4 x 4 can be modelled by having four sets of trains, each of four pink rods (assuming 1 pink is 4 whites). The arrangement of the rods is optional. They can all be stretched out into one long train, or a super - rectangle shape made showing 4 sub - rectangles each of four pinks. Alternatively, a pink cube can be constructed. For any of these the standard whole number name can be figured out by whatever means seem appropriate to the students doing it, by measuring, by reference to the multiplication table or by mentally coming to the standard answer, 64.

If students are familiar with the manipulation of cross symbols, a yellow - yellow cross for example may be called '5 to the 2th power', equivalent to a train of 5 yellows and, therefore, to 25 whites.

With another yellow rod across the topmost rod of the cross, we can read '5 times 5 times 5', all in an example of a *tower,* accept that it is also called '5 to the third power' and invent many other one - color towers.

Because a cross is directly associated with a rectangle (a *square* rectangle in the special cases of one - color crosses) and that with a train, we are able to transform any tower into a corresponding train for measurement if the standard whole number name is wanted. Should we begin, for example, with 3^8 the bottom cross of 3 x 3 can have substituted a length of 9. So $3^8 = 9 \times 3^6$. By continuing like this and successively changing a cross into a train the tower gets lower and lower, the bottom train getting longer and longer. This can be recorded thus:

$$3^8 = 9 \times 3^6$$
$$= 27 \times 3^5$$
$$= 81 \times 3^4$$
$$= 243 \times 3^3$$
$$= 729 \times 3^2$$
$$= 2187 \times 3^1$$
$$= \underline{6561}$$

Relation Between Exponential and Standard Whole Number Forms

At first there may not appear to be much difference between 2^3 and 3^2 between 2^8 and 8^2, or between 10^6 and 6^{10} ... and so on.

In each pair there are the same number names, the order is reversed and maybe one of them is written slightly smaller than the other. We are so used to that not making a difference that in order to have a different impact as we read, some greater involvement is probably necessary. One way is to consider the equivalent whole number forms for each.

Ex. Which is greater? 2^3 or 3^2?
2^8 or 8^2?
10^6 or 6^{10}?
Change each to a whole number and compare.

Ex. What general rule can be stated, if any, about the relative sizes of a^b and b^a? Investigate examples when a is greater than b, or a = b.

Examples of exponential notation of particular interest may be those which indicate very large numbers, like the size of atoms or the number of seconds in a lifetime.

e.g. seconds in 2 weeks approximately, 1×10^6 (1 million)

1 year	31×10^6
an average lifetime	2×10^9
the age of the Earth	1×10^{17}
the age of the Universe	3×10^{17}

The size of things compared to the diameter of an atomic nucleus:

a speck of dust or the thickness of a hair	25×10^9
Diameter of the Earth	144×10^{19}
Diameter of our Solar system	144×10^{25}

From such we can well understand the wisdom of using exponential forms and calling it the 'scientific notation'. Hopefully, we can see that all of the other notation is just as scientific!

Equivalents

Many equivalents of exponential forms are also available and fascinating to invent, regardless of how the basic principles were first introduced.

5^9 for example is certainly equivalent to $5^7 \times 5^2$, $5^6 \times 5^3$, $5^5 \times 5^4$, $5^4 \times 5^5$, $5^3 \times 5^6$, $5^2 \times 5^7$, because the sum of each pair of exponents is 9. Proceeding

$$3^8 \div 3^1$$

$$3^{10} \div 3^1 \div 3^2$$

$$3^4 \times 3^1 \times 3^2$$

$$3^1 \times 3^1 \times 3^1 \times 3^1 \times 3^1 \times 3^1 \times 3^1$$

$$3^9 \div 3^2$$

$$3^2 \times 3^5$$

$$3^{07} \times 3^{100}$$

$$3^{9010} \div 3^{9003}$$

$$3^{7009} \div 3^{7002}$$

$$3^{100020} \div 3^{100013}$$

$$3^4 \times 3^1 \times 3^2$$

$$3^{11,111} \times 3^{-11,104}$$

$$3^{12,111} \times 3^{-12,104}$$

Names or equivalents of 3 ^{7}

further in the pattern we might expect 5^8 x 5^1 is another equivalent, although 5^1 can hardly be seen as a 'repeated multiplication'. It seems reasonable, however, to accept this as another name for 5 and in fact is defined this way in mathematics.

What of 5^9 and 5^0? $9 + 0$ is 9 but what meaning can 5^0 carry? If we wish to have 5^9 x $5^0 = 5^9$ then clearly 5^0 must be accepted as an equivalent of 1. That too is fixed by definition; it fits into the pattern of the exponents in multiplicative forms.

Ex.   What is the standard whole number form for, 6^0, 8^0, n^0?

Ex.   Invent some products, each equivalent to 5^9, the terms being in exponent form, perhaps with one x sign, perhaps with two or more.

Ex.   Repeat the previous exercise beginning with *any* exponent form.

Another development occurs when we see that 5^6, being equal to 5^3 x 5^3, is also equivalent to '5 to the 3rd power *all of it* raised to the 2nd power'. This leads to more:

$$(5^2)^3 \qquad (5^6)^1 \qquad (5^1)^6$$

Other examples however may produce a richer crop, because the exponents have many factors:

$$5^{36} \qquad (5^6)^6 \qquad (5^9)^4 \qquad ((5^3)^3)^4 \qquad ((5^4)^3)^3 \quad \dots \text{ and so on.}$$

Ex. Begin with 7^{60} and write non - standard but exponential equivalents. Does anyone want to work out its whole number equivalent?

Ex. If 5^0, 8^0, 23^0 ... and so on, are each equivalent to 1, what is the standard exponent form for:

$$((3^0)^{75})^{27} \qquad (((17^3)^{21})^0)^{54}$$

Write others like these.

Fractional Exponents

Since $9 = 2\frac{1}{2} + 6\frac{1}{2}$ could it be that $5^9 = 5^{2\frac{1}{2}} \times 5^{6\frac{1}{2}}$?   We can harldly find a tower of rods equivalent to $5^{2\frac{1}{2}}$ nor get a clump of clumps if a clump represents $5\frac{1}{2}$. Whatever meaning can be attached to a number with a fractional exponent no application can therefore be made to the measurement of lengths or to the subdivision of counters in this way. However without worrying for the moment about applications it seems that if we wish to use $5^{2\frac{1}{2}}$, for example, it will be an equivalent of a number between 5^2 and 5^3, between 25 and 125 that is.

Moreover, the patterns for the multiplication of exponent forms do fit in all cases, for if the exponents are fractions the multiplication of the exponent forms is equivalent to another exponent form whose exponent is the sum of the separate exponents, and a sum of two fractions is always possible. Many other equivalents for 5^9 (or any other) using fractional exponents can therefore be generated:

$$5^{\frac{1}{2}} \times 5^{8\frac{1}{2}} \quad 5^{\frac{1}{4}} \times 5^{\frac{1}{4}} \times 5^{8\frac{1}{2}} \quad 5^{\frac{1}{2}} \times 5^{\frac{1}{2}} \times 5^{\frac{1}{2}} \times 5^{\frac{1}{2}} \times 5^7$$

Ex. Use fractional exponents to get a rich crop of equivalents for, say, 4^8. Or any other example.

Negative Exponents

If people have extended the multiplication table to include negative entries it may have occurred to some that negative exponents might be tried. Would not $5^{10} \times 5^{-1}$ be an equivalent to 5^9 since $10 + {}^-1 = 9$? The matter seems compelling and gives us another trigger for the generation of equivalents, all in exponent form.

Ex. Generate exponent equivalents of 8^9, or use any other example, the exponents being a mixture of positive and negative integers.

Ex. Find the standard exponent form for each of these:

$$b^2 x b^7 x b^{-3} x b^{-5} \quad (x^{-3})^2 \times (x^5)^0 \times x^6 \quad (((a^{10})^{-2})\tfrac{1}{2})^{\frac{1}{5}}$$

Alternatively, the following sequence suggests a number of patterns:

$$a^4 = a \times a \times a \times a$$
$$a^3 = a \times a \times a$$
$$a^2 = a \times a$$
$$a^1 = a$$

On the left the exponents are ... 4, 3, 2, 1. On the right the number of terms in the product form is successively 4, 3, 2, 1, though the last is not strictly in product form.

a x a x a x a is transformed into a x a x a by dividing by a.'The same transformation operates on a x a x a to produce a x a and again 'dividing by a ' gives the next righthand name a . Maintaining these patterns further than a^1 = a, the results would seem to be:

$$a^0 = a \div a = 1$$

$$a^{-1} = 1 \div a = \frac{1}{a} \text{ or} \frac{1}{a^1}$$

$$a^{-2} = \frac{1}{a} \div a = \frac{1}{a^2}$$

$$a^{-3} = \frac{1}{a^3} \ldots \text{ and so on.}$$

Division

With stress on multiplication of exponents no mention has been made of division! Clearly however, as soon as $5^4 \times 5^5$ is accepted as equivalent to 5^9 we could have posited that,

$$5^9 \div 5^4 = 5^5$$

$$\text{and } 5^9 \div 5^5 = 5^4$$

The pattern, such that with multiplication of exponent forms in the same base the separate exponents are added to give the standard form of the product, infers the companion pattern that when exponent forms are *divided* the exponents are *subtracted* to give the standard form.

Division can be defined thus, as the inverse of multiplication, analogous to what was experienced with whole number multiplication and division. With rods if a tower of yellow rods, for example, represents a number in exponent form it is multiplied by another yellow by placing such a rod across on top of the tower. Inversely, to divide a tower by one of its rods it would be *taken away,* decreasing the height of the tower by one rod.

The use of positive and negative and fractional exponents will apply to division just as they did for multiplication, save only that exponents will be subtracted rather than added.

Ex. Generate division names in exponent forms equivalent to 8^9, or to x^{10}, or to any other example. Have some using one division sign, others with more than one, still others with a mixture of x and ÷.

Square Roots and Reciprocals

Although in a preliminary study of exponents it may be considered unnecessary to go further than already suggested, there are important deductions which follow from what has been intuitively understood so far,

a) There are special forms for a^1:

$$a^{1/2} \times a^{1/2} = a^1$$

But $a^{1/2} \times a^{1/2}$ also equals $(a^{1/2})^2$, 'a to the ½, squared'.

Consequently $a^{1/2}$ must be the *square root* of a^1, since $a^{1/2}$ is that number which when multiplied by itself gives a^1.

$$a^{1/2} = \sqrt{a^1}$$

Further,

$$a^{1/3} \times a^{1/3} \times a^{1/3} = a^1$$

$a^{1/3}$ is the cube root of a^1, also written $a^{1/3} = \sqrt[3]{a}$

Continuing, $a^{1/4}$ is the fourth root of a,
$a^{1/5}$ is the fifth root of a . . . and so on.

Ex.   What are the standard names for,

$$25^{1/2} \qquad 64^{1/3} \qquad 100^{1/2} \qquad \left(\frac{36}{81}\right)^{1/2}$$

Ex.   Express x $3/2$ in square root form.

Extend the pattern to any number, x say, with a fractional component.

b) Another special form, for a^0, is the like of:

$$a^{+3} \times a^{-3} = a^{+3 + -3} = a^0 = 1$$

$$\text{So } a^{+3} \times a^{-3} = 1$$

$$\therefore a^{+3} = \frac{1}{a^{-3}} \text{ and } a^{-3} = \frac{1}{a^{+3}}$$

$$\text{Similarly, } a^{+4} = \frac{1}{a^{-4}}, \qquad a^{+5} = \frac{1}{a^{-5}} \quad \cdots \quad a^{+n} = \frac{1}{a^{-n}} \quad \cdots$$

and so on.

A number in exponent form, whose exponent is an integer, is the reciprocal of the corresponding form whose exponent is the opposite integer.

Ex.   What are the standard whole number or rational number forms for:

$$4^{-2} \qquad \frac{1}{3^{-1}} \qquad 10^{-3} \qquad 1^{-1}$$

Using fractional integer exponents, what are other forms for:

$$25^{-1/2} \quad 8^{-1/3} \quad 8^{+1/3} \qquad 27^{-1/3} \quad 16^{+1/4} \quad 64^{-2/3}$$

$$1000000^{+3/2} \qquad\qquad 1000000^{+2/3}$$

Mathematics

Finally for exponents in this book and relating back to the mathematics of chapter 1 it is worth noting that we have here a good example of the system called an 'abstract group'. There have been others, though not identified as such.

For if the system consists of the set of elements:

$$a^0, a^{+1/2}, a^{+2/2}, a^{+3/2}, \ldots, a^{+1/3}, a^{+2/3}, a^{+4/3} \ldots$$
$$a^{-1/2}, a^{-2/2}, a^{-3/2}, \ldots a^{-1/3}, a^{-2/3}, a^{-4/3} \ldots$$

with the operation of multiplication as defined, then we may put to the test each of the four requirements for an abstract group:

1) <u>Closure</u>   The system is closed because the product of any two elements gives an element of the set;

2) <u>Associative</u>   This principle holds good because it rests upon the associativity of the *sums* of the exponents. They *are* associative being positive, negative or zero fractional *sums*.

3) <u>Neutral element</u> a^0 is the neutral element for, $a^0 \times a^n = a^n$ whatever positive or negative integer n represents.

4) <u>Inverses</u>   Every element of the set has an inverse:
$a^{+1/2} \times a^{-1/2} = a^0 = 1,$ $a^{+m/n} \times a^{-m/n} = a^0 = 1,$ for all m, n as whole numbers, n not 0.

Thus the exponent forms constitute an abstract group under the operation of multiplication. Consequently they have the same structure as the mathematical system inherent in our story and challenges of 'My Three Sons!'

17

The Problem of Problems

In school, writings like $1 + 2 = \square$, $4x^2 + x + 1 = 0$, $\frac{25.62}{8.1}$ are called problems. Sometimes it is a student who is a problem. At other times the problem may well be to understand what the teacher is talking about!

Then there are 'word problems', formerly and delightfully called 'promiscuous problems' in school texts of the 1900's — which only means problems of various kinds! They are paragraphs describing applications of arithmetic and mathematics to contexts of money, measurement, speeds, comparisons of various quantities . . . and so on, everyday examples met in business, industry, commerce and the home. The hope is that students, by tackling them in this form, will learn to tease out the numbers from the rest of the writing, operate on them correctly and appropriately get the right answer.

All these meanings given to the word 'problem' fit with the Webster dictionary definition:

> - from the Greek, *proballein,* to throw forward
> - a question proposed for solution or consideration
> - a question, matter, situation, or person that is perplexing and difficult
> - in maths, anything required to be done.

Traditionally in Arithmetic, however, it is the *word* problems which seem to attract the 'perplexing and difficult' interpretation. Remarks that students may be able to do the rest of the arithmetic but they 'can't do word problems' are continually heard.

We assert that the approach suggested in this book, throughout all topics, has been one which has indeed required that things be done, that questions have been regularly proposed for consideration by the learners. This is necessary for all mathematics learning. Word problems, therefore, become only particular instances of a generalized problem approach in which patterns are always

190

sought, determinations have to be made about strategies and decisions have to be made as to what is relevant, what not.

As already suggested (chapter 5) any arithmetical statement can be applied to contexts of money, measurement or other 'real - life' situations. It would be wise, therefore, for students to become accustomed to such applications by experiencing what it means to make up such applications from the beginning of their study, to realize what an enormous range of applications can be met, rather than wait to meet those made up by other people in a textbook chapter headed 'Problems'. They can do this by inventing 'word problems' for themselves as well as interpreting examples written by others.

The nature of possible challenges are illustrated by the following:

$3 + 4 = 7$ Write a story about objects in a store which uses this addition problem.

$\frac{3}{4} + \frac{1}{2} = 1\frac{1}{4}$ Say how this could apply to telling time from a clock.

$\frac{2}{3} \times \frac{5}{8} = \frac{5}{12}$ Describe how this multiplication of fractions might be about dividing up some areas of land.

.15 of 7 = 1.05   What could this have to do with a tip paid for a restaurant meal?

8% of 23   Say how this could apply to an inflation increase in the cost of food.

$^-3 + {}^+2 + {}^+1$ Could -3 represent goals scored by opponents in a hockey match? If so what could the complete number name represent?

Variations are without limit, subject only to the ingenuity of learners.

With such experience of 'applied mathematics' students will come to textbook problems as just other people's examples of what they themselves can do. They will not be topics to avoid, topics which are harder than those continually experienced with the 'pure mathematics'. If learners have continually dealt with generalizations of number relationships, then to clothe them in other language and real - life contexts, they will be more ready than usual to reverse the strategy: sift out the important relationships from the enveloping garb of language and context, and operate on them appropriately to the specific situation described.

Other direct aids to tackling word problems include,

1.   Cross out words which are not important to the basic problem.
2.   Extract from the complete description what *is* important, possibly in short forms with numerals, operational and relational signs.
3.   Construct a model in rods, blocks, string, paper . . . and so on, to represent the context.
4.   Make drawings or diagrams.
5.   In inventing one's own word problems sometimes insert lots of irrelevant words and number names. Trade with the inventions of other students and solve their problems.

6. Repeat the strategy of No. 5 trying purposely to obfuscate a reader, burying the information relevant to the required solution, within a mass of irrelevancies!

Conclusion

There are many problems to the learning and teaching of mathematics. One of them is *not* that some people can learn the subject while others cannot. Our first three chapters of this book constitute an attempt to convince all who read that there is real mathematics for everyone.

The problem is not that mathematics is not enjoyable. It is a great joy when every individual knows the creative possibilities by firsthand experience. Its usefulness after that is an added and important by - product.

The problem is not even that educators do not know what needs to be done for important mathematics to be mastered by everyone. That is now known.

The references in the bibliography indicate a growing literature reflecting this relatively new assertion.

The only unsolved problem is how to get the majority of teaching adults to know that mathematics *is* for everyone and acquiring the know - how of reflecting this with children of any age. This book is a contribution towards the solution of that!

Suggestions for further reading:

Cuisenaire, G. and Gattegno, C. *Numbers in Colour*, Heinemann, Toronto, 1954.

Dienes, Z. P. *The Power of Mathematics*, Hutchinson, London, 1963.

Gattegno, C. *What We Owe Children*. Outerbridge and Dienstfrey, New York, 1970.

————. *Teaching foreign language in schools* – the Silent Way, 1963.

————. *Teaching Reading with Words in Colour*, 1968.

————. *Now Johnny can do Arithmetic*, 1971.

————. *The Common Sense of Teaching Mathematics*. 1974, Educational Solutions Inc., New York.

Ginsburg, H. *Children's Arithmetic – the Learning Process*, Van Nostrand Co., Toronto, 1977.

Goutard, M. *Mathematics and Children*. Cuisenaire Company of America, New Rochelle, 1964.

Moskowski, B. A. "The Acquisition of Language," *Scientific American*. November, 1978.

Skemp, R. R. *The Psychology of Learning Mathematics*, Penguin Books Ltd., England, 1971.

Trivett, J. V. *Mathematical Awareness*, I and II, 1963.

————. *Games Children Play for Learning Mathematics*, 1974, Cuisenaire Co. of America, New Rochelle.

————. "The End of the 3 R's" – the Beginning of a New Language Art", *Mathematics Teaching*. Association of Mathematics Teachers, England, September. 1978.